Resilient Architecture Design for Voltage Variation

Synthesis Lectures on Computer Architecture

Editor
Mark D. Hill, *University of Wisconsin, Madison*

Synthesis Lectures on Computer Architecture publishes 50- to 100-page publications on topics pertaining to the science and art of designing, analyzing, selecting and interconnecting hardware components to create computers that meet functional, performance and cost goals. The scope will largely follow the purview of premier computer architecture conferences, such as ISCA, HPCA, MICRO, and ASPLOS.

The Datacenter as a Computer: An Introduction to the Design of Warehouse-Scale Machines
Luiz André Barroso and Urs Hölzle
2009

Computer Architecture Techniques for Power-Efficiency
Stefanos Kaxiras and Margaret Martonosi
2008

Chip Multiprocessor Architecture: Techniques to Improve Throughput and Latency
Kunle Olukotun, Lance Hammond, and James Laudon
2007

Transactional Memory
James R. Larus and Ravi Rajwar
2006

Quantum Computing for Computer Architects
Tzvetan S. Metodi and Frederic T. Chong
2006

Resilient Architecture Design for Voltage Variation

Vijay Janapa Reddi and Meeta Sharma Gupta

ISBN: 978-3-031-00611-1   paperback
ISBN: 978-3-031-01739-1   ebook

DOI 10.1007/978-3-031-01739-1

A Publication in the Springer series
SYNTHESIS LECTURES ON ADVANCES IN AUTOMOTIVE TECHNOLOGY

Lecture #22
Series Editor: Mark D. Hill, *University of Wisconsin, Madison*
Series ISSN
Synthesis Lectures on Computer Architecture
Print 1935-3235   Electronic 1935-3243

Resilient Architecture Design for Voltage Variation

Vijay Janapa Reddi
The University of Texas at Austin

Meeta Sharma Gupta
IBM T.J. Watson Research

SYNTHESIS LECTURES ON COMPUTER ARCHITECTURE #22

ABSTRACT

Shrinking feature size and diminishing supply voltage are making circuits sensitive to supply voltage fluctuations within the microprocessor, caused by normal workload activity changes. If left unattended, voltage fluctuations can lead to timing violations or even transistor lifetime issues that degrade processor robustness. Mechanisms that learn to tolerate, avoid, and eliminate voltage fluctuations based on program and microarchitectural events can help steer the processor clear of danger, thus enabling tighter voltage margins that improve performance or lower power consumption. We describe the problem of voltage variation and the factors that influence this variation during processor design and operation. We also describe a variety of runtime hardware and software mitigation techniques that either tolerate, avoid, and/or eliminate voltage violations. We hope processor architects will find the information useful since tolerance, avoidance, and elimination are generalizable constructs that can serve as a basis for addressing other reliability challenges as well.

KEYWORDS

voltage noise, voltage smoothing, $\frac{di}{dt}$, inductive noise, voltage emergencies, error detection, error correction, error recovery, transient errors, power supply noise, power delivery networks

Contents

Preface

Over the past decade, designers have grappled with the challenge of building efficient systems with respect to power, performance, and cost. Of these, power has emerged as a first-order obstacle. Looking into the future, the next major challenge for us is to build robust and reliable systems that meet historically established reliability standards without compromising energy-efficiency or the "price-to-performance" ratio. Unfortunately, operating a system close to an efficient design point makes the system susceptible to unreliability, whereas allowing large safety margins makes the system inefficient. Understanding and harnessing this interplay between reliability, power, and performance is the crucial next step to building sustainable computer systems in the future that meet next-generation systems requirements.

In the future, we will require better integration and collaboration between hardware and software. As technology trends force us to build for typical-case design, error-detection and recovery mechanisms will become pervasive, and as such, we must identify and develop new machine organizations that are capable of dynamically detecting and recovering from errors in the field across all layers of the computing stack, including computer architecture, system software, and applications. The benefits of such a collaborative computing stack are twofold. First, it eliminates performance and energy inefficiencies that arise at each layer from maintaining strict abstraction between hardware and software. Second, it eliminates power and area overheads that arise from the use of circuit- and microarchitectural-level techniques that mitigate the various sources of errors.

As per this book, the general idea of a resilient processor architecture is to rely on hardware techniques for fail-safe execution, but, when possible, attempt to optimize away recurring sources of errors via software techniques. Such an approach avoids recurring hardware penalties that arise from the hardware's limited view of execution activity using software which has much more global knowledge about the processor's execution activity.

Vijay Janapa Reddi and Meeta Sharma Gupta
May 2013

Acknowledgments

The work described in this book is the result of many researchers' hard work, insights and creativity. We owe credit to all of them. In addition, we would like to specifically thank David Brooks and Gu-Yeon Wei from Harvard University for their numerous creative and insightful research ideas and discussions on hardware and software co-design for voltage variation. We would also like to thank Mark Hill from the University of Wisconsin-Madison and Michael Morgan from Morgan & Claypool for giving us this opportunity to survey and describe state of the art literature on this topic. With their help, we were fortunate to receive detailed and very useful feedback on an earlier draft version of the book. We are thankful to the reviewers who provided us with constructive feedback.

Vijay Janapa Reddi and Meeta Sharma Gupta
May 2013

CHAPTER 1

Introduction

Continued advancement of complementary metal-oxide-semiconductor (CMOS) technologies provides the well-known benefits of device scaling. However, as feature sizes shrink and chip designers attempt to reduce supply voltage to meet power targets in large multicore systems, parameter variations are becoming a serious problem. Parameter variations can be broadly classified as device variations incurred due to imperfections in the manufacturing process and environmental variations due to fluctuations in on-die temperature and supply voltage. Collectively, these parameter variations greatly affect the speed of circuits in a chip; delay paths may slow down or speed up due to these variations. The traditional approach to dealing with parameter variations has been to overdesign the processor based on the most pessimistic operating conditions to allow for worst-case variations. As the gap between nominal and worst-case operating conditions in modern microprocessor designs grows, the inefficiencies of worst-case design are too large to ignore. Recognizing the inefficiencies in such a design style, researchers have begun to propose architecture-level solutions that address worst-case conditions.

In this chapter, we first introduce the problem of parameter variations. We begin by giving a broad but brief overview of the various fundamental issues to set the context for the reader. We explain process, thermal, and voltage variations and quantify their impact on the worst-case design margin for processor robustness, which motivates us to consider processor designs for the typical case mode of operation. Given the large breadth of work, it is not feasible to comprehensively address all parameter variation solutions in one book. Therefore, we narrow the scope of our discussion in this book to voltage variation, which is seen as a major impediment to building robust and energy-efficient systems in the future.

1.1 PARAMETER VARIATIONS

Parameter variations began to pose a major design challenge for CPU designers in technologies starting from the 65 nm process [15]. These parameter variations can be broadly classified into two categories: static variations and dynamic variations (Figure 1.1). Static variations or process variations are caused by the inability to precisely control the fabrication process at small-feature technologies. It is a combination of systematic effects [14] (e.g., lithographic lens aberrations) and random effects [91] (e.g., dopant density fluctuations). On the other hand, dynamic variations are caused by the applications' runtime characteristics and can be further classified into voltage variations and temperature variations. Voltage variations can be caused by IR drops (i.e., product of current I passing through resistance R) in the supply distribution network. Alternatively, voltage variations

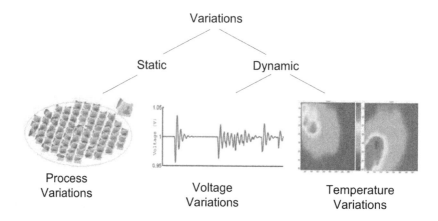

Figure 1.1: Parameter Variations. Illustration of the different types of variations: process, voltage, and temperature variations.

are also caused by $L\frac{di}{dt}$ noise (i.e., product of inductance and the rate of change of the current flowing through it over time) under varying load conditions. Temperature variation is caused by spatially- and temporally-varying factors. Parameter variations affect the worst-case delay of critical paths and hence, directly affect performance and power. Variations also introduce spread in path delays across a chip both spatially (across different units in a chip) and temporally (as workload behavior causes voltage and temperature to fluctuate). All of these variations are becoming more severe and harder to tolerate as technology scales to smaller feature sizes.

Designers typically account for parameter variations by inserting conservative margins that guard against *worst-case* variation characteristics to guarantee the system's functional correctness under all operating conditions. The size of these timing margins depends on how designers account for the impact of variations. Conservatively, designers may treat each source of variation indepen- dently and determine worst-case margins by simply summing the required margin for each source separately. This conservative approach ignores important interactions that can exacerbate variation effects. For instance, there are substantial amounts of spatial, temporal, and application-level slack designers can(not) exploit. Spatial slack occurs when variations delay different units of a processor by different amounts. Temporal slack is exposed due to the runtime behavior of workloads. Application- level slack emerges because of differences among workloads. On the basis of experimental data, we will discuss these various forms of slack further in the following paragraphs.

Spatial Slack   Systematic variation in process parameters, such as threshold voltage, gate length, and oxide thickness, create differences across the core. Different software phases load different functional units to a greater or lesser degree. For example, the fixed-point unit (FXU) in the simulated POWER6 processor has both high power density and high activity fluctuations, leading to larger temperature spikes and deeper voltage droops than those of the floating-point unit (FPU).

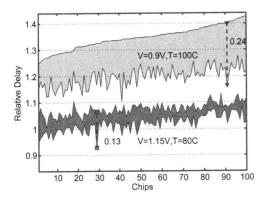

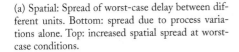

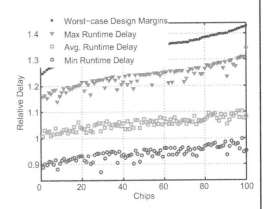

(a) Spatial: Spread of worst-case delay between different units. Bottom: spread due to process variations alone. Top: increased spatial spread at worst-case conditions.

(b) Temporal: Spread of delay due to run-time behavior. The maximum, average, and minimum delay over all 16 benchmarks and 5 units per chip (a total of 80 points/chip).

Figure 1.2: Spatial and Temporal Slack. Differences across processor units and between application phases the designers can exploit.

Figure 1.1 shows the difference between the worst-case circuit delay for different units under two different operating conditions, a nominal operating condition of 1.15 V at 80°C and a worst-case design point of 0.9 V and 100°C. Under nominal operating conditions a maximum slack of 13% relative delay is available. The slack increases to 24% when the operating conditions are based on the worst-case design points (though for a different chip). Variation in process parameters changes the voltage-delay or temperature-delay relationships of circuits, widening the gap between units as conditions worsen.

Temporal Slack   Figure 1.1 shows the maximum run-time delay across the SPEC CPU2006 program suite for 100 simulated chips (generated with process variations as described in Chapter 2), along with the delay for worst-case design margins (V=0.9V, T=100C). There is a slack of 10% for nearly every chip between the worst-case delay and the maximum runtime delay. The average run-time delay shows an even greater amount of temporal slack. It is approximately 20% across 100 different chips.

Application-Level Slack   Unlike process variations, voltage and temperature variations are closely coupled to workload characteristics, which vary from application to application and from phase to phase within an application. Figure 1.1 plots the variation of delay across major units for different programs. Within a program, the units have different delay profiles and the relative order of units differs between programs. Figure 1.1 shows the variation of *omne*, a SPEC CPU2006 program,

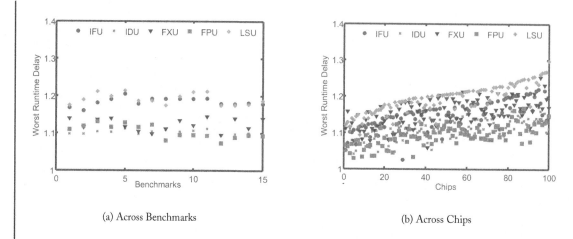

(a) Across Benchmarks

(b) Across Chips

Figure 1.3: Worst runtime circuit delay variations. (a) Different applications for a single chip and (b) Running the SPEC CPU2006 program *omne* with different chips.

across different chips. The load-store unit (LSU) has the worst behavior for a majority of the chips, owing to large swings in activity followed by long stall periods. The variation in the relative order of units between different chips is primarily the result of within-die process variations.

In summary, it is important to consider process-voltage-temperature (PVT) variations in combination and not as independent entities. Moreover, it is important to consider both spatial and temporal differences across the processor die, along with within and across applications behavior. Section 1.2 discusses this issue and motivates the need for design solutions that consider all sources of variation simultaneously. Section 1.3 motivates the need for solutions to accommodate the large gap between worst-case and nominal operating conditions.

1.2   WORST-CASE DESIGN

All sources of parameter variation lead to timing overhead and uncertainty, but each kind of variation has different characteristics. Process variations, which result from imperfections in the manufacturing process, are potentially large variations in device features such as the threshold voltage and gate length of fabricated transistors, with both systematic and random components. Voltage variations are closely related to workload-dependent activity fluctuations on the order of tens to hundreds of cycles. These fluctuations cause dips in the power-supply distribution network owing to the interaction of the activity patterns and parasitics of the supply network. Similarly, although at a much coarser time scale, variations in activity also induce temperature differences across the chip.

Figure 1.4 illustrates the timing margins required when considering isolated and combined sources of variations. These timing margins are set by the worst-case delay path within a chip. The

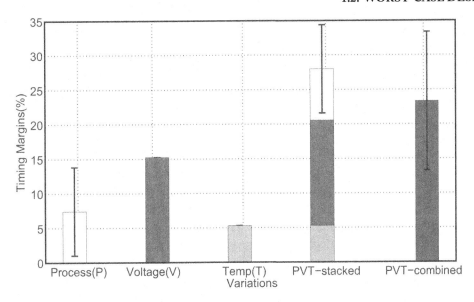

Figure 1.4: Impact of PVT Variations on Timing Margins. Simple *stacking* leads to a larger mean and smaller spread in required timing margins as compared to the *combined* effect.

process parameters, gate length, threshold voltage, nominal supply voltage (V_{nom} = 1.15V), and nominal temperature (T_{nom} = 80 °C) are based on a 65 nm technology node and ITRS specifications [47]. The first three bars (process, voltage, and temperature) represent the timing margins required for each of the variations when considered in isolation. We evaluate process variation across a batch of 100 chips by modeling and simulating both systematic and random effects.

The bars represent the mean timing margin across 100 chips. The error bars represent the maximum and minimum timing margins observed across the 100 chips. Voltage variations are evaluated by running the suite of programs on a nominal chip with no process variation and at T_{nom}. The bar represents the mean of the worst-case timing margin required for the suite of programs considered, indicating a 15% timing margin required for handling voltage variations. Temperature variations are evaluated by determining the timing margins required for a chip with no process variation and at a temperature of 100 °C operating at V_{nom}. Temperature variations require 5% margins to ensure correctness in case of worst-case temperature across the core.

Simply stacking the individual margins together (PVT-stacked) results in a simple, but conservative, approach to setting worst-case timing margins. In contrast, the figure also shows the resulting timing margins from simulating all sources of variations together. When PVT variation effects are combined (PVT-combined), the average margin required is reduced, but the spread between maximum and minimum margins is increased. This larger spread in margins primarily results from the interaction between voltage and process variations. Faster chips, consisting of transistors with lower threshold voltages, are less sensitive to voltage droops (sudden drops in voltage) and can operate with tighter margins. On the other hand, the transistors with higher threshold voltages in slower chips are more sensitive to voltage droop and require larger margins. Hence, the spread between maximum and minimum margins increases. The reduction in the average margin can be attributed to runtime temperatures, typically being lower than applying a worst-case 100 °C penalty. The slowest chip, with the highest threshold voltages across the chip, exhibits lower leakage power to ameliorate thermal effects and slightly reduce the maximum required margin.

Binning   Given that simply stacking margins misses important interactions found by considering PVT variations together, designers devise solutions that address all sources of variations in combination and not as individual, orthogonal components. A typical outcome of this process is the functional operating frequency of the chip. Speed grading (or binning) is used to determine the maximum functional operating frequency of a chip. Speed-grading or binning is an industry standard approach that refers to the process used by manufactures to determine the maximum functional operating frequency of a chip, so that it can be offered to customers at various appropriate speed-grades. Typically, all chip manufacturer offers chip at a range of different clock frequencies. Assuming the range of clock frequencies across 600, 733, 800, 867, 933 MHz and 1 GHz, two chips at 975 MHz and 991 MHz will be speed-(down)graded into the 933 Mhz lot. Similarly, speed grading will bin a 1067 MHz or 1022 MHz chip into the 1 GHz lot. It is the statistical nature of variation from chip to chip and lot to lot that primarily lead to the wide variation in operational frequencies.

Manufacturers use functional test patterns to determine the various speed grades. A set of functional test patterns is developed over the course of design development and productization using a variety of methods. For example, designers hand-generate custom functional tests that are guided by critical timing paths. Some test cases are automatically generated on the basis of timing verification tests while others are extracted from application programs to represent real-world execution scenarios. During the initial stages, there are also exercises that involve "speed-path hunting." It is a tedious but important step.

It is important to note that conventional uniprocessor binning strategies, which sort chips according to the maximum operating frequency, may apply inadequately toward multicore processors because within-die variations are becoming significant. Applying the minimum of the maximum safe operating frequencies for various cores of a chip multiprocessor (CMP), which is commonly referred to as the *min-max* metric, tends to severely penalize CMP cores that could otherwise sustain higher clock frequencies by themselves. A real-life example of frequency variation in silicon was observed in an 80-core Intel test chip [25]. At a supply voltage of 1.2 V, there was a 28% variation between the fastest core's frequency of 7.3 GHz and slowest core's frequency of 5.7 GHz. The gap grew at lower voltages to 59% when the nominal voltage was set to 0.8V. Thus, min-max sacrifices substantial performance to achieve simplicity by running all cores at the frequency of the slowest core.

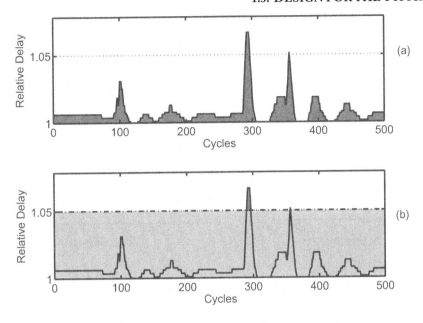

Figure 1.5: Performance Loss due to Timing Margins. The figure depicts an example highlighting the performance loss incurred by providing timing margins (bottom) compared to ideal cycle-by-cycle frequency tracking (top).

1.3   DESIGN FOR THE TYPICAL CASE

Conservative designs operate at the worst-case timing margin, ensuring robustness, but with performance loss due to lower processor frequency. Because worst-case conditions can be severe, but infrequent, operating with conservative worst-case margins is costly. This section explores the widening gap between nominal and worst-case conditions, assuming that infrequent worst-case scenarios can be handled by a *fail-safe* mechanism.

Figure 1.5 shows a snapshot of delay over 500 cycles for SPEC CPU2006 program *h2ref* run on a chip with no process or temperature variations. The delay variation is solely attributed to voltage droops. There are infrequent large droops in voltage causing occasional increases in delay. To estimate the benefits of design for the nominal case, we consider an ideal cycle-by-cycle frequency tracking scenario that adjusts clock frequency according to the cycle-by-cycle delay (akin to an asynchronous design). We illustrate the performance loss of such a system as the area under the curve in Figure 1.5 (top). We compare such a scenario to one that assumes fixed timing margins and that handles worst-case conditions with a fail-safe mechanism. A fail-safe mechanism can recover the system correctly in presence of violations. Figure 1.5 (bottom) illustrates this example, which applies a timing margin of 5%, but incurs an equivalent fixed-performance penalty plus additional performance penalties for any margin violations.

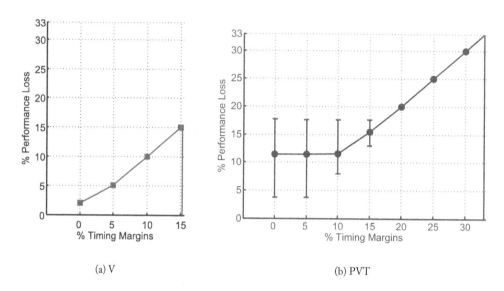

(a) V

(b) PVT

Figure 1.6: Need for Typical Case Design. Performance loss at various timing margins is depicted for (a) voltage variations and (b) process-voltage-temperature variations.

Figure 1.6 compares the performance loss of the ideal cycle-by-cycle tracking scheme (represented by 0% margins) with that of applying fixed timing margins while simulating the entire program at the 65 nm technology node. Figure 1.6(a) plots the performance loss corresponding to different timing margins in the presence of voltage variations. An ideal asynchronous design incurs 2% performance loss due to delay fluctuations. As timing margins increase up to the worst-case 15% level, performance loss can mostly be attributed to the fixed margins as delay fluctuations are infrequent. Figure 1.6(b) presents a similar plot that considers the impact of PVT variations. The error bars again correspond to the spread, due to process variations for 100 simulated chips. With 0% timing margins, the fastest chip exhibits a 4% loss in performance and the slowest chip incurs a much larger penalty of ∼17%. Applying larger fixed-timing margins penalizes the fastest chips until the fastest and slowest chips suffer similar losses beyond 20%, where the margins again dictate performance loss. The plot extends to 33%, which represents a scenario that sets margins with respect to the worst-case delay across all chips. Thus, this analysis shows that the severe but infrequent nature of worst-case runtime conditions motivates architectural design strategies that avoid running with timing margins based on worst-case conditions. Such strategies have the potential to recapture up to 20% of performance loss on average.

1.4   SCOPE OF THE BOOK

Answering the above questions requires careful analysis and introspection. In this book, we offer an in-depth look at how to develop a resilient processor architecture to address voltage variation, which is frequently also known as voltage or inductive noise, or the $L\frac{di}{dt}$ effect. Of all variations, we chose voltage variation as the specific example because it necessitate the large guardbands (see Figure 1.4) and thus, is a major source of processor inefficiency.

We describe a cross-layer approach for mitigating voltage variation with the goal of eliminating the penalties to power, performance, and cost that arise in the use of any circuit techniques and microarchitectural changes for mitigation. As technology trends force us to build for typical-case design, error-detection and recovery mechanisms will become pervasive. To sustain continued increases in system performance, we must identify and develop new machine organizations that are capable of dynamically detecting and recovering from errors in the field across all layers of the computing stack, including computer architecture and system software. The benefits are twofold: (1) this eliminates performance and energy inefficiencies that arise at each layer from maintaining strict abstraction between hardware and software and (2) it eliminates power and area overheads that arise from the use of circuit- and microarchitectural-level techniques that mitigate the various sources of failures. A cross-layer solution will rely on hardware for fail-safe execution, but, when it is possible, it will attempt to optimize away recurring sources of errors via software techniques.

The rest of this book is structured as follows. First, we provide the reader with background on voltage variation in Chapter 2. We focus on the basics of modeling voltage variations at the architecture level. Since modeling involves trade-offs between speed and accuracy, we discuss the trade-offs between various modeling approaches.

Power-delivery subsystem design affects voltage variation within a processor. We discuss this in Chapter 3. Most of our effort is targeted toward low-cost robust solutions for voltage variations, therefore, we give a detailed characterization of the design of a power-delivery subsystem and its impact on the rest of the system. We cover a wide range of power-delivery subsystems for modern processors and show how different characteristics affect the occurrence of timing-margin violations.

In Chapter 4, we examine how traditional solutions cope with voltage variation. Several of these solutions work at the expense of power, performance, and cost. As such, we propose and forecast a forward-looking, resilient architecture design for coping with voltage variations systematically and more efficiently.

Chapter 5 summarizes a variety of solutions that push toward typical-case design by tolerating emergencies. The goal here is to allows margin violations to occur, but when they do, the architecture has the ability to roll back to a guaranteed-correct processor state.

However, in the event that tolerating emergencies is prohibitively costly, as in the case of a costly rollback mechanism, it is more advantageous to proactively anticipate margin violations based on heuristics or logic and take precautionary measures that avoid the issue altogether by dynamically reacting to the impending violation. In Chapter 6 we cover a spectrum of avoidance techniques that have been proposed to predict voltage emergencies.

While hardware-based solutions can be effective, they are often reactive. They sense, detect, and respond repeatedly, even when the activity can be smoothed-away by potentially simple instruction and thread scheduling techniques. Software has a global view of execution, and as such, it can perform such transformations, albeit at a higher penalty. However, this penalty can be amortized if the optimization is effective and the program runs long-enough to reap the benefits of the optimization. Thus, it is possible to fall back on hardware for immediate reaction (albeit suboptimal) to emergencies (e.g., aging protection circuits) and rely on software to eliminate repeated stress-related occurrences, thereby eliminating "waste" and improving overall system efficiency. In Chapter 7, we provide motivation for the software-based solutions and evaluate the trade-offs, specifically in the context of multicore systems where building hardware-based solutions is substantially more challenging than building software-based solutions.

Finally, in Chapter 8 we provide our thoughts on promising new directions for resilient architecture design and deployment.

CHAPTER 2

Modeling Voltage Variation

Efforts to address microprocessor power dissipation through aggressive supply voltage scaling and power management require that designers be increasingly cognizant of power supply variations. These variations, due primarily to fast changes in supply current, can be attributed to architectural gating activity and microarchitectural events that reduce power dissipation. In order to study this problem, coarse- and fine-grained parameterizable models for power-delivery networks are required that enable system designers to study localized and global on-chip supply fluctuations in processors. In this chapter, we focus on the different voltage modeling approaches and discuss the associated modeling tradeoffs.

2.1 A QUICK PRIMER

Sudden current swings due to activity fluctuations in a microprocessor, when coupled with parasitic resistances and inductances in the power-delivery subsystem, can give rise to large voltage swings. Equation 2.1 highlights how current fluctuations interact with the impedance of the system (represented by Z). A decrease in supply voltage leads to an increase in the delay of the gates. Voltage also impacts both the system's dynamic power and leakage power.

$$V = V_{DD} - V_{Drop} \tag{2.1a}$$
$$V_{Drop} = Z \times I_{Instantaneous} \tag{2.1b}$$
$$I_{Instantaneous} = P_{Leakage} + P_{Dynamic} \times V_{DD} \tag{2.1c}$$

In order to prevent sudden current changes over a large range of frequencies (typically from the kilohertz range to the processor's operating clock frequency) from becoming voltage spikes, designers carefully design the processor's power supply such that it has a low impedance over a wide range of frequencies. To achieve such a low impedance power supply, several strategies are used that include some combination of decoupling capacitors and voltage regulators. Circuit designers build the system with on-die capacitors, on-package capacitors, and voltage regulators in combination with off-package capacitors and regulators. The decoupling capacitors strive to compensate for impedance that is introduced by the parasitic inductance of the power-supply network at each level of the power-delivery hierarchy.

We will present the detailed aspects of voltage modeling later, in Section 2.2. Several factors must be considered, such as the voltage regular module (VRM), motherboard, package, and off-chip decoupling capacitors. Moreover, there are different approaches to modeling voltage variation: the distributed grid model, the sparse grid model, and the impulse-response-based model. The trade-offs in accuracy and simulation speed vary.

2.2   THE POWER-DELIVERY NETWORK (PDN) SUBSYSTEM

Voltage variations are strongly coupled to the characteristics of the underlying power-delivery subsystem. Therefore, it is important to have good models for processor activity, power consumption, and the power-delivery subsystem. In this section, we begin with an overview of the power-delivery components that need to be included in any voltage model. Then we describe three different methods of modeling voltage variation: a detailed grid model, an impulse-response-based model, and a sparse grid model.

Figure 2.1 gives an overview of the components to model for voltage variation, including the VRM, motherboard, package, and off-chip decoupling capacitors. The off-chip power-delivery network includes the motherboard, package, and off-chip decoupling capacitors and parasitic inductances, which are modeled via a ladder RLC network. It is also necessary to model the bulk capacitance on the PC board and the package as an effective capacitance and effective series resistance. Voltage regulator modules (VRM) typically have response frequencies in the sub-MHz range, which is much lower than the challenging higher frequencies associated with the entire power-delivery network. For simplicity, the power supply is typically modeled as a fixed-voltage source, which is scaled with respect to the average current draw to deliver V_{DD} at the bump nodes, mimicking the feedback loop associated with the VRM. The electrical properties of these network elements determine what impact current variations at different frequencies have on chip-level voltage.

For any given power-delivery network, the impedance profile of the power-delivery subsystem describes the distribution network's frequency sensitivity. An example impedance plot of a Pentium 4 processor is shown in Figure 2.2. The mid-frequency resonance peak is the dominant property responsible for the inductive noise problem. The mid-frequency resonance peak shown at 100 MHz in Figure 2.2 is mainly due to the interaction of the package inductance and decoupling capacitance [13]. It is not easy to compensate for the inductance of the wires between the die and the package. This inductance often causes a peak of high impedance [11, 43] in the power supply at the resonance of the chip capacitance and the package inductance.

"Noise" at this resonant frequency, which is typically in the range of 10–100 MHz [11], is considered the most dangerous since it can be excited by an instruction loop whose repetition rate coincides with this resonant frequency and contains a large current step. The net effect is a transient voltage droop large enough to exceed the margin Although some extreme circuit techniques for compensating for this exposed inductance exist, such as increased on-die capacitors and on-die voltage regulators [41], these techniques are prohibitively expensive and are not broadly applicable for general-purpose mainstream computing where cost is an influential design parameter for system architects.

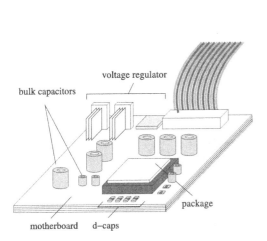

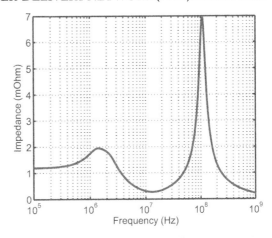

Figure 2.1: Power-Delivery Subsystem. This figure depicts the main components associated with a processor's power-delivery subsystem.

Figure 2.2: Example Impedance Plot. This plot depicts an impedance plot modeled for a Pentium 4 package, with mid-frequency resonance at 100 MHz and a peak impedance of 7mΩ.

2.2.1   DISTRIBUTED GRID MODEL

We first present a detailed yet flexible power-delivery model that captures the characteristic mid-frequency resonance, transients related to board and package interfaces, and localized on-chip voltage variations. As we show, this model provides good accuracy but at a simulation cost, which makes it appropriate for feedback-driven experiments where a circuit solver interfaces with architectural-level simulation for a few 1000s of cycles as opposed to longer simulations of millions of cycles.

Figure 2.3 presents a detailed model of the power-delivery network with a distributed on-chip power-supply grid. Figure 2.4 illustrates the distributed on-chip grid model used in our analysis. We assume a flip chip package design. The package connects to the chip through discrete controlled collapse chip connection (C4) bumps. The C4 bumps are modeled as parallel connections (via RL pairs) that connect the grid to the off-chip network, with each grid point having a bump connection. The decision to model a flip chip package is made because it is the frequent design choice for high-performance processors. This model could be easily adapted to capture the behavior of a bond wire package instead.

The on-chip grid itself is modeled as an RL network. The evenly distributed on-chip capacitance between the V_{DD} and GND grids is modeled in two ways—C_{spc} represents the decoupling capacitance placed in the free space between functional units, and C_{blk} represents the intrinsic parasitic capacitance of the functional units. In contrast, an on-chip lumped model would consist of a single RLC network connected across the package-to-chip interface. When scaling the size of the grid, the values in the RLC network are scaled proportionally. The amount R and L is scaled by the change in length of each grid segment, and the total capacitance between the power and ground

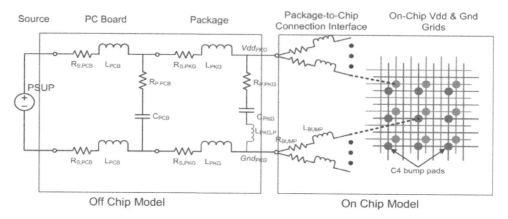

Figure 2.3: Model of a Power-Delivery System. The figure shows the parasitics of the package, the package-to-chip interface, and the on-chip grid.

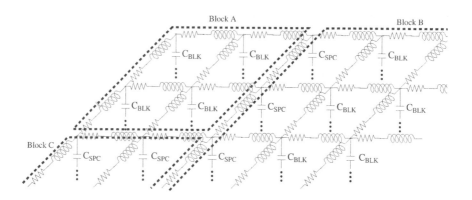

Figure 2.4: On-Die Distributed Grid Model. Illustration of the parasitics of the on-chip grid model, including the additional decoupling capacitance on the processor core.

planes is redivided among the grid nodes. This yields a closely matching impulse response as the number of grid nodes is increased.

Table 2.1 provides the values of the resistances, inductances, and capacitances used for the PCB and package and on the die for the power-delivery model. These values were chosen to match the measured off-chip impedance of the Pentium 4 processor [7, 45]. It is important to note that these parameters can easily be modified to model different architectures and power-delivery networks. This parameterizable distributed grid model was used in analyzing voltage variations in the context of next-generation chip-multiprocessor (CMP) architectures using both real applications and synthetic current traces [37].

Table 2.1: Parameters for the Power-Delivery Model. The values of various resistances, inductances, and capacitances in the power-delivery subsystem. The values represent a power-delivery model such as a Pentium 4 processor power-delivery subsystem

Resistance	Value	Inductance	Value	Capacitance	Value
$R_{pcb,s}$	0.094 mΩ	L_{pcb}	21 pH	C_{pcb}	240 μF
$R_{pcb,p}$	0.1666 mΩ				
$R_{pkg,s}$	1 mΩ	L_{pkg}	120 pH	C_{pkg}	26 μF
$R_{pkg,p}$	0.5415 mΩ	$L_{pkg,p}$	5.61 pH		
$R_{bump,grid}$	40 mΩ	$L_{bump,grid}$	72 pH		
				$C_{decoupl}$	335 nF
R_{grid}	50 mΩ	L_{grid}	5.6 fH	C_{blk}	0.12nF
				C_{spc}	1.5nF

Table 2.2: Simulation Speed of Using a Distributed Grid Model. This table presents the simulation times of various grid sizes (100K cycles)

Grid Size	1x1	2x2	4x4	6x6	8x8	12x12	16x16	20x20	24x24
Speed (s)	5	28	174	493	1070	1423	1830	4620	7280

Accuracy of the power-delivery model is strongly dependent on the grid resolution of the on-chip distributed grid. A finer distributed grid should be able to give a finer sampling of the voltage variations across the die, and hence, provide a more accurate depiction of the variation across the chip. A coarser resolution of the grid fails to capture the finer variations across the chip. However, simulation speed is a critical issue if localized power-delivery models must interface with architectural power and performance simulators. Table 2.2 outlines the speed of the lumped model and various grid models for a 100K-cycle sample run with a detailed power/performance model for a high-performance microprocessor.

2.2.2   IMPULSE-RESPONSE-BASED MODEL

While a given microprocessor's PDN is a complex system consisting of several different components (e.g., voltage regulator module, package, on-die capacitors, etc.) [7, 27], a simplified second-order lumped model [42, 87] can adequately capture its resonance characteristics with impedance peaking in the mid-frequency range of 50 to 200 MHz and can be reasonably modeled as an underdamped second-order linear system [42] as described by Equation 2.2.

$$a\frac{d^2}{dt^2}y(t) + b\frac{d}{dt}y(t) + cy(t) = f(t) \tag{2.2}$$

Ideally, the supply voltage across a processor should be constant. However, due to dynamic current fluctuations and the non-zero impedance of the PDN, large voltage fluctuations can occur. One way to characterize voltage variations is by convolving the microprocessor's instantaneous current profile with the impulse response of the PDN (Equation 2.3).

$$v(t) = i(t) * h(t) \tag{2.3}$$

Most of the earlier work that seeks to address voltage variation at the architectural level uses a simplified second-order lumped model [42, 87], which captures the system's mid-frequency response. However, such models fail to capture within-die spatial variation of voltage. While detailed grid models have also been proposed and used, the large number of nodes lead to prohibitively high simulation times; hence, we discuss an intermediate model next.

2.2.3   SPARSE GRID MODEL

A detailed distributed grid model suffers from slow simulation time, of the order of few hours per million instructions, making it impossible to use such a model for detailed workload characterization. To deal with the model's simulation-time complexity, a simpler and faster on-chip model can be used that retains the per-unit or per-block characteristics. The proposed model's simulation time is approximated at 1 hr/100M instructions, which is orders of magnitude faster than the detailed grid model.

By appropriately scaling R, L, and C of each unit's power grid with respect to area, the model can enable relatively fast simulations while maintaining the accuracy that closely matches a detailed grid model. The R, L, and C for any unit must be scaled as given in Equations 2.4, where A_{unit} represents the area of the unit and A_{core} represents the area of the core. R_l, L_l, and C_l correspond to those values that are found in a lumped model.

$$R = R_l * \frac{A_{core}}{A_{unit}} \tag{2.4a}$$

$$L = L_l * \frac{A_{core}}{A_{unit}} \tag{2.4b}$$

$$C = C_l * \frac{A_{unit}}{A_{core}} \tag{2.4c}$$

To demonstrate confidence in the sparse model, the sparse model is validated against a 12x12 distributed on-chip grid model. The sparse model used for validation is a simplified grid model with a single point for each unit. Figure 2.5(a) shows the mean and variance of the error in the voltage reported by the abstract model as compared to a detailed model. The mean of the error in voltages is

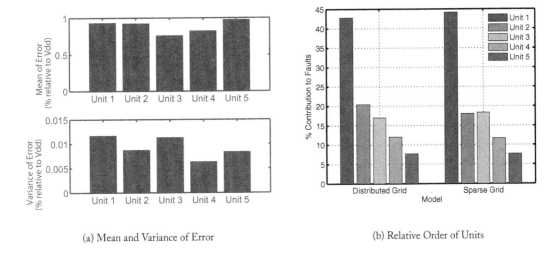

(a) Mean and Variance of Error   (b) Relative Order of Units

Figure 2.5: Accuracy of the Sparse Grid Model. (a) Shows the mean and variance of the difference of voltage reported between the sparse and detailed grid model, reported each cycle, relative to the nominal power supply. (b) Depicts the relative order of the faults across the different units modeled in the core for a timing-margin of 6%.

less than 20mV, which translates to less than 1% of error in the voltage for a 1.15 V nominal power supply.

Additionally, the relative order of the per-unit behavior also reflects the model's good accuracy. Figure 2.5(b) shows the descending order of the units for voltage violations at a 6% timing-margin averaged across all the programs. The analysis indicated that the test programs spend a significant amount of time around 6% timing-margin and hence, a 6% timing-margin is applied as the reference margin. The relative order of the units is maintained in the abstract model as compared to the detailed distributed model, with little difference in the absolute distribution of the faults for each unit.

2.3   SUMMARY

Accurate and fast models are important to study and understand the impact of parameter variations on future processor architectures. In this chapter, we briefly explored process and temperature models, alluding to the works that provide greater details. We then focused on voltage variation and described three fundamental voltage modeling approaches. There is a trade-off to be made between the accuracy and speed of the three voltage models. The choice depends on the type of study being conducted. For architectural studies, speed is often a concern, because simulating millions of clock cycles is a substantially slow and time-consuming process that can render the approach inapplicable. The sparse grid model, however, achieves satisfactory performance without sacrificing modeling accuracy.

CHAPTER 3

Understanding the Characteristics of Voltage Variation

Voltage variation is closely linked to the detailed characteristics of the power-delivery network (PDN) subsystem. In this chapter, we characterize the interaction between current consumption profiles and the PDN that leads to voltage variation using the impulse-response-based voltage model described in Chapter 2. We also explore how different package parameters affect the magnitude of the variation for a given processor. It is important to understand these interactions in order to design a robust resilient processor architecture for handling voltage variation.

3.1 CURRENT PULSES

Sudden short spikes in current can cause voltage variations, but the magnitude of the variation is largely determined by the amount of charge built up over a specific time interval. These current spikes inside the processor are typically caused by power management techniques, such as clock gating and power gating, which have become increasingly important to stay within the peak power ceiling. By carefully pruning the clock tree and disabling portions of inactive circuits, designers effectively reduce the dynamic switching power. However, the sudden change of switching current may introduce undesirable voltage variation.

Figure 3.1(a) shows the voltage transients for current pulses of varying amplitudes and durations. The first and second pulses have the same width, but, the second pulse has a higher amplitude. A sufficiently high amplitude can induce violations. While the last two pulses shown have large integrated charge, they do not cause significant variations in core voltage. This indicates that isolated pulses with a certain amplitude/width combination can lead to voltage variation.

Even if the voltage fluctuation caused by a current pulse in isolation does not exceed voltage margins, a series of such pulses at the PDN's resonance frequency may lead to a voltage emergency. Figure 3.1(b) shows the voltage response for a series of current pulses. The first sequence of current pulses has a period of 30 cycles, which corresponds to a frequency of 100 MHz for a 3 GHz processor. If the resonance of the PDN also occurs at 100 MHz, voltage swings gradually build up and exceed emergency thresholds. Thus, it is important to consider both isolated pulses and resonating pulses

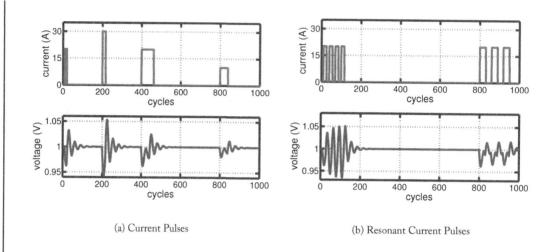

(a) Current Pulses

(b) Resonant Current Pulses

Figure 3.1: Current Pulses and Effect on Voltage Transients. The plot depicts isolated current pulses and resonating current pulses with a period of 100 MHz and the corresponding voltage behavior.

when designing an architectural solution to handle voltage emergencies. In general, there are a few basic types of current profiles commonly found in real-world programs. These include the following.

1. *Step Current:* This type of current profile commonly occurs when a core suddenly changes state. Figure 3.2(a) shows a sudden increase/decrease in activity. This can occur, for example, when the firmware enables sleep/active state transitions that power up/down cores. Typically, these step currents are very large in magnitude.

2. *Pulse Current:* These are caused by sudden, short increases/decreases in the core's activity, which are typically caused by long stalls in the processor. Figure 3.2(b) shows an example of isolated pulses with a certain pulse width.

3. *Resonating Current:* Periodic behavior largely associated with recurring activity patterns generally attributed to loops in an application. In particular, a periodic sequence of current pulses occurring at or near the power-delivery network's resonant frequency is typically of most interest because it leads to large voltage variation. These resonating currents are shown in Figure 3.2(c) for *bzip2*.

Given the observed application profiles, we can perform in-depth analysis by substituting synthetic current profiles in order to interrogate the power-delivery network for a wide range of problematic scenarios. In the following paragraphs, we focus our efforts on describing the effects of step currents and sequences of pulse currents on the power-delivery network that lead to voltage variations. Current pulses of long enough duration and magnitude can be classified as step currents. In simulations, the worst-case analysis is achieved by using two states for each core: max-power and

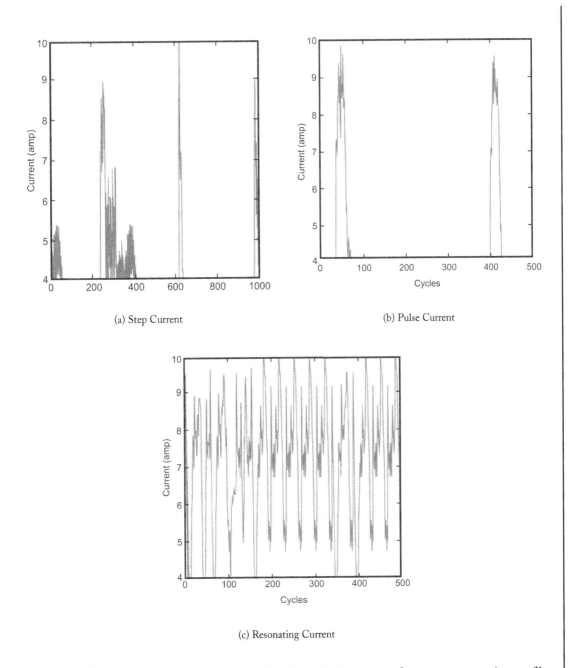

(a) Step Current

(b) Pulse Current

(c) Resonating Current

Figure 3.2: Typical current pulse characteristics. Three distinct types of current consumption profiles from real-world programs.

min-power. The maximum power state refers to when a core is drawing maximum power, which for the purposes of the following text corresponds to 10W/core in simulations of a four-core processor. The minimum power state refers to the core consuming minimum power from the system, which corresponds to 4W/core. In the remaining analysis, we model steps and pulses with these max/min power levels to mimic powering up/down cores or activities observed in the SPEC programs.

Step Current Effects   Current steps can induce large voltage fluctuation around the nominal voltage. Large step currents are typically observed whenever cores are powered on/off. Thus, voltage variation is an alarming scenario during system boot-up/reset. Figure 3.3 shows the voltage variation on the chip when all four cores are powered on simultaneously. Given that a step is composed of signals across a wide range of frequencies, the initial drop in voltage and the subsequent ringing can be attributed to the power-delivery network's high-frequency resonance (100 MHz). The voltage dip that occurs at 500 cycles can be attributed to the low-frequency resonance. The voltage eventually stabilizes to the system's nominal voltage (1V).

Figure 3.3 (inset) plots the minimum voltage with respect to the number of simultaneously engaged cores. As expected, the worst-case voltage drop is observed when all the cores are switched on simultaneously (i.e., four cores vs. one, two, and three cores). To avoid such a worst-case condition, designers typically use a simple staggering mechanism to gradually ramp the current profile with assistance from the firmware or microcode [81]. The cores are turned on one by one in some order to avoid the sudden in-rush of current.

Periodic Pulse Current Effects   Figure 3.4 plots the peak voltage swing observed across the chip when the current consumption of all four cores simultaneously switches between maximum and minimum power at different frequencies with a 50% duty cycle. This leads to a resonating current, which, in effect, is a periodic current pulse occurring at frequencies within the resonant band of the power-delivery network.

As anticipated and shown by the impedance plot of the power-delivery network, the worst-case voltage swing occurs around 100 MHz (i.e., the package's resonant frequency). Given resonating currents, the voltage ripple initially grows and then settles to a periodic waveform around the nominal voltage (as shown in Figure 3.4 (inset)). In steady state, even small current pulses can induce large peak-to-peak swings; thus, they are of further interest.

It is useful to categorize resonating current pulses into the following two important categories: *Locally Resonant*, where each core individually has periodic current pulses at the resonant frequency; and *Globally Resonant*, where the aggregate current, seen globally across the die, has, or appears to have, current pulses at the resonant frequency of the power-delivery network's impedance. Let us investigate the combination and interaction of these two types of resonating currents.

1. *Locally Resonant and Globally Resonant:* This represents a scenario where each core has resonating current and the combined (or average) current pulses across all of the cores are also at the same resonant frequency. Voltage swings grow as the number of resonating cores increases

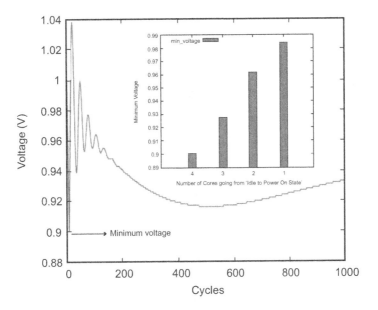

Figure 3.3: Step current effect. Powering on cores together is bad for voltage variation.

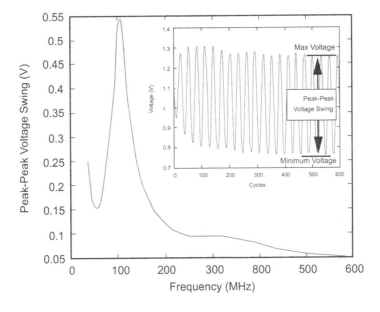

Figure 3.4: Periodic currents effect. A periodic current pulse occurring at frequencies within the resonant band of the power-delivery network can cause large peak-to-peak voltage swings.

due to the higher aggregate current amplitudes (i.e., constructive interference). The theoretical worst-case condition occurs when current pulses across all of the cores are aligned in phase.

2. *Locally Resonant but Globally Nonresonant* — Locally, the cores are resonating, but due to phase differences, the combined view seen by the system is not a resonating wave. For conditions where the resonating currents across all cores are phase-shifted with respect to one another, currents between the cores can interact to cancel out some of the effects of the locally resonating currents at the global scale. When 50% duty cycle current pulses are 90° out of phase with one another, as shown in Figure 3.5(a) (left), the currents combine to appear as a constant amplitude current, globally.

Figure 3.5(a) (right) presents the case where resonating current pulses are each offset by 60°. In this case, the combined currents have periodicity at the resonant frequency, but the stepwise waveform leads to smaller voltage fluctuations. Figure 3.5(b) summarizes the effect of varying the phase shift between resonant currents across the four cores and a range of duty cycles, on the resulting peak-to-peak voltage swing magnitudes seen across the chip. As seen before, the worst-case condition is when all current pulses are aligned in phase (0 or 360).

Generally, a larger duty cycle means higher overall current draw and, hence, it leads to larger voltage variation. Interestingly, in the simulated four-core processor example, interactions between cores lead to the most canceling when current pulses are phase-shifted by multiples of 90°. Given this dependence on the number of cores, a 16-core processor may exhibit similar dips for phase differences occurring in multiples of 22.5°.

3. *Locally Nonresonant but Globally Resonant:* The previous conditions were examples of resonating currents occurring locally within the cores. But, we must also consider the scenario where each core does not consume currents that pulse at the resonant frequency; rather, as shown in Figure 3.6(a), the combined waveforms look like a resonating current. Given a tightly coupled power-supply grid with low impedance connections between the cores, Figure 3.6(b) (left) shows that each core internally experiences resonant voltage variation. In fact, there is little difference to the condition where the combined current waveform is evenly distributed across the four cores, whose resulting voltage waveforms are plotted in Figure 3.6(b) (right). The only difference is the higher local ripples that occur according to the local current pulses.

Hence, simply studying current pulses at the resonant frequency at the individual core level will not be sufficient to fully characterize voltage variation at the global scale of a multicore chip. The example we discussed further emphasizes the need to understand and study intercore interactions at the various levels of the system and design process, from application-derived current profiles to the low-level power-delivery network itself.

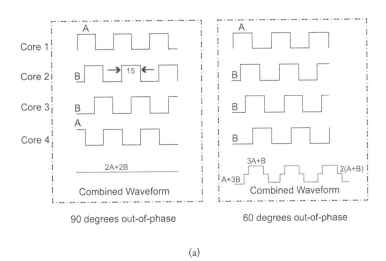

90 degrees out-of-phase   60 degrees out-of-phase

(a)

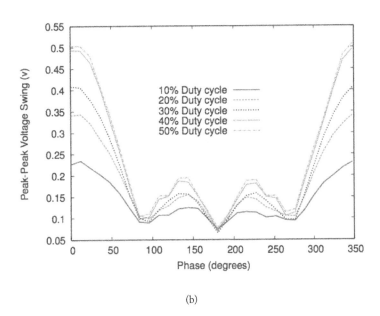

(b)

Figure 3.5: Locally Non-Resonant but Globally Non-Resonant. (a) Effect of number of resonating cores on the peak voltage swing and (b) Effect of phase difference on the peak voltage swings.

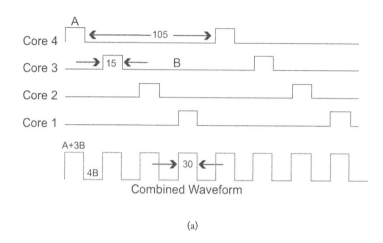

(a)

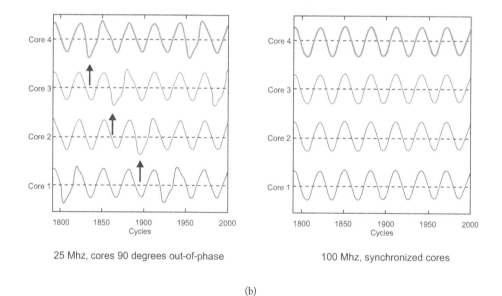

(b)

Figure 3.6: Locally Non-Resonant but Globally Resonant. (a) Example of a locally non-resonant but globally resonant input and (b) Snapshot of voltages for the four cores.

3.2   PDN CHARACTIERSTICS

Given that a power-delivery subsystem can be modeled as a second-order linear system, the package model's response to current variations is largely governed by three factors: Q (quality factor), C (resonance cycles), and Z (peak impedance). These factors affect the robustness and correctness of any solution for handling timing-margin violations. In this section, we analyze the effects of these three factors on voltage emergencies.

Quality Factor (Q)   A system's quality factor is the ratio of the resonant frequency to the rate at which it dissipates its energy. This factor determines the width of the resonance, or the *resonance band* of the system. A higher Q leads to a greater buildup of voltage for currents oscillating within the resonance band. Q depends on the effective inductance (L) and resistive losses (R) at the resonant frequency ($f = \frac{1}{2*\pi*\sqrt{(LC)}}$) as shown in the following equation:

$$Q = \frac{2*\pi*f*L}{R} \ .$$

(3.1)

A good package will have lower parasitic inductance and hence, lower Q than a poor package. Figure 3.7(a) shows different packages with different Q, highlighting that higher-Q packages have a narrower resonance band and higher impedance at the resonant frequency. Higher impedance means that applications with current variations within the resonance band experience larger voltage fluctuations. Figure 3.7(b) illustrates how different packages with different Q factors affect voltage emergencies on a subset of the SPEC CPU2000 programs. As Q increases, the fraction of cycles where the voltage extends beyond ± 5% thresholds increases for all programs. However, the slope for each program differs, with *crafty* experiencing the steepest increase in timing-margin violations. This can be attributed to the programs' differing current profiles.

The PDN Q factor defines an important constraint on any technique designed to handle voltage emergencies. Specifically, the rate of change of voltage will depend on Q. Later, in the next few chapters, we will discuss the implication of the rate of change of voltage on the design of architectural solutions to deal with voltage emergencies.

Resonance Cycles (C)   This factor represents the number of processor cycles corresponding to one period of the PDN resonant frequency. As processor frequency increases, while the PDN resonant frequency remains fixed, C also increases. For example, a resonant frequency of 100 MHz for a 10 GHz processor would result in a C of 100 [72], whereas C would be 30 for a 3 GHz processor.

Voltage emergencies strongly depend on this resonance cycles metric. Figure 3.8 plots the resulting voltage fluctuations for three settings of C and shows that the minimum width of an emergency-inducing isolated current pulse differs for different resonance cycles. In fact, this width depends on the resonant frequency of the PDN such that a larger C tends to require wider current pulse widths, in terms of the number of processor cycles.

Figure 3.9 shows how the fraction of cycles with timing-margin violations varies with processor frequency for a given package. The package considered here has a resonant frequency of 100 MHz.

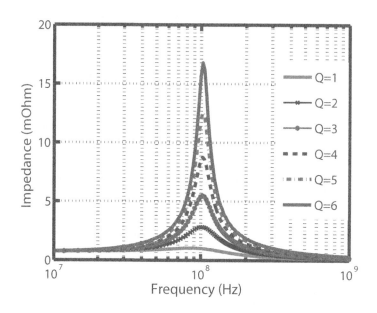

(a) Packages with different Q factors.

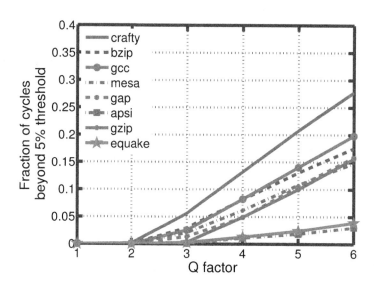

(b) Variation of cycles in voltage emergencies with Q

Figure 3.7: Sensitivity to Q. A package with a higher Q-factor would lead to more timing-margin violations (i.e., voltage emergencies).

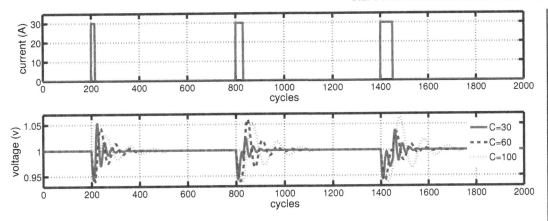

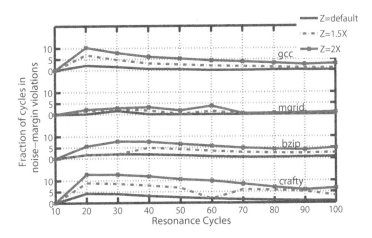

Figure 3.8: Sensitivity to Resonance Cycles. The plot shows an example of an isolated pulse for different resonance cycles metric. A higher resonance cycle metric requires a wider pulse to cause violations (crossing −5%).

Figure 3.9: Resonance Cycles and Peak Impedance (Z). The figure depicts the sensitivity to resonance cycles metric and peak impedance(Z) for some CPU2000 programs.

We can see that the peaks for the different programs shown here are at lower values of the resonance cycles metric (20-30 cycles). This is because, in these programs, current pulses tend to have smaller widths—both for resonating and isolated pulses.

Peak Impedance (Z)   This factor represents the peak impedance for the power-delivery subsystem at its resonant frequency. Ideally, this peak (or target) impedance should be as low as possible to avoid voltage emergencies. However, efforts to reduce this peak impedance can increase system cost. It would require improved package design, etc. Therefore, circuit and architecture designs must cope with higher-than-desired impedance to avoid voltage emergencies. Figure 3.9 shows that, as the package's peak impedance increases, the timing-margin violations also increase across all applications and resonance cycles.

The extent of voltage variation varies across different power-delivery subsystem designs, and is also closely related to the current consumption profiles. It is important to guarantee that the mechanism used for handling voltage emergencies is robust across a wide range of package and processor characteristics.

3.3   MICROARCHITECTURAL EVENTS

Thus far, we have explored voltage variation in response to stimulant current pulses and their interaction with the underlying PDN. In a real processor, these current pulses are caused by execution stalls, which are the result of microarchitectural activity. Thus, in this section, we study events that cause stalls, such as branch mispredictions and cache misses.

Gupta et al. [38] considered several microarchitectural parameters that could affect voltage variation, such as the number of entries in the reorder buffer, the instruction fetch queue, and the load/store queue, along with microarchitectural events such as cache misses and pipeline flushes. On the basis of their findings, we describe the perturbation effects of microarchitectural events on processor activity using real program examples and show that they can lead to voltage emergencies. We also discuss patterns in activity that let us not only identify voltage variation points uniquely, but also predict their recurring occurrences [38, 40, 78].

Figure 3.10 shows a snapshot of pipeline activity for the topmost loop in *equake*. Pipeline statistics, such as occupancy of reorder buffer and commit rate, are depicted along with microarchitectural events that were tracked. Several other microarchitectural parameters, such as number of entries in the ROB, instruction fetch queue, and load/store queue, were also considered but are deemed to be not as useful.

In the figure, the presence of a long stall due to an L2 cache miss can be observed (shown by marker point 1). During the time it takes to service the L2 miss, pipeline activity ramps down, as seen in the current profile. However, after the L2 miss data is available, functional units become busy and there is a sudden increase in activity (shown by marker point 2). This steep increase in current leads to a significant voltage drop (shown by marker point 3). Other events such as TLB misses, L1 misses, or flushes are not present in the pipeline during this window, which clearly suggests that the L2 miss in this code region caused the voltage fluctuation in *equake*. We now categorize such events into six distinct categories.

Memory Miss Events   Stalls can occur in the pipeline owing to loads missing the L1 cache, and larger stalls occur when the loads miss the L2 caches. The large miss penalty associated with L2 or higher cache misses can drain the active instructions in the pipeline and can result in long periods of inactivity. It is important to note that an L2 miss that results in a pipeline stall may lead to an emergency, while an L2 miss that does not stall the pipeline likely will not. This period of inactivity

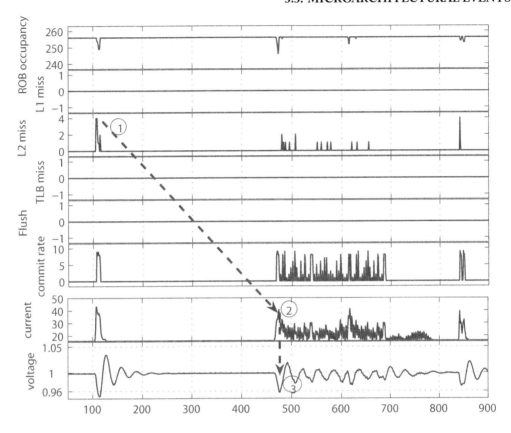

Figure 3.10: Memory Miss Events. Example of a microarchitectural stall effect on voltage from procedure smvp in program *equake*.

following a miss is characterized by low current draw (as shown in 3.10). A sudden increase in activity happens when the L2 miss returns, leading to execution of all dependent instructions. These bursts of activity following a long period of inactivity must be avoided.

Pipeline Flush Events Misprediction of branches leads to flushing the entire pipeline when the branch is resolved. This leads to a sudden decrease in activity following a flush event; however, a few cycles later, activity ramps up because of noncontrol instructions at the branch target. If the period of low and high current draw matches coincides with the periodicity of the package characteristics, resonance buildup of voltage occurs. Figure 3.11 shows one such example of a snapshot of *art*. The L2 miss (shown by marker point 1) is responsible for the initial inactivity and subsequent increase in activity when it returns, leading to a drop in voltage, but the subsequent pipeline flushes (shown by markers 2, 3, 4) occur periodically, leading to further voltage drops. The snapshot also shows the presence of L1 misses, but their effect has been determined to be small on voltage drop [38].

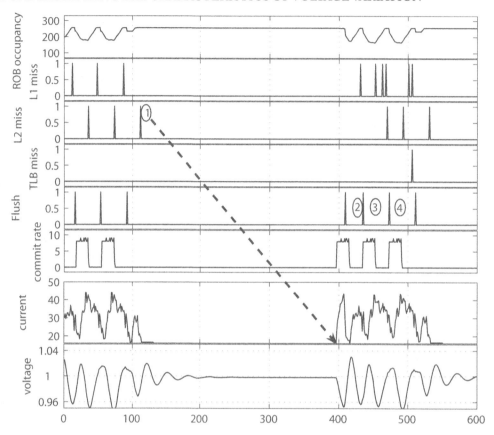

Figure 3.11: Pipeline Flush Events. Example of memory miss events and resonant flushes on voltage from procedure match in program art.

Long Latency Events A long chain of dependent floating-point operations, such as divides, can lead to a long wait period for dependent instructions leading to a stalling effect of the pipeline. Figure 3.12 shows one such example. The occurrence of the long-latency divide operation is shown at the top of Figure 3.12. The divide instruction halts all processor activity due to a lack of instruction-level parallelism (ILP), and thus, the processor's issue rate drops, as does the current draw. However, when the divide instruction finishes execution, there is a sudden burst of activity. Instructions dependent on the divide are ready to execute, leading to a dramatic increase in ILP. Rapidly issuing instructions causes the current to suddenly increase, which in turn causes the voltage to drop quickly (15 cycles).

Uncharacterized Certain applications or phases of application execution display an an even more complex interplay of events. In such cases, it is sometimes difficult to identify a single cause or a set of interacting events that lead to voltage swings. Figure 3.13 shows one such example with high

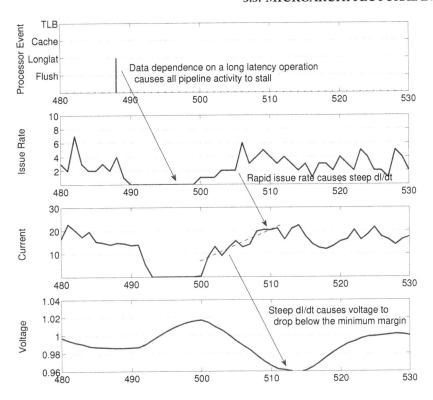

Figure 3.12: Long Latency Event. Example of a long latency stall effect on voltage in program *Sieve* from the JavaGrande benchmarking suite.

frequency noise occurring around the nominal voltage. We see L1, L2, and pipeline flush events all occurring close to one another. In such cases, it is difficult to clearly identify a single dominant microarchitectural cause for the emergency.

Mixed Events   Whether a microarchitectural event at a particular location causes an emergency may depend on activity just before and after the event. Even a small loop like that in Figure 3.14(b), extracted from the *gcc* program of SPEC CPU2006, can have behavior phases with markedly different activity patterns. Figure 3.14(a) is a snapshot of activity within that loop over 880 cycles. It shows three repeating phases of the loop. Phase A uses paths $1 \rightarrow 4$ and $1 \rightarrow 2 \rightarrow 4$, while phase B uses only path $1 \rightarrow 2 \rightarrow 3 \rightarrow 4$. The issue rate of phase A is relatively low, while that of phase B is quite high. The flush events labeled 1 are caused by branch mispredictions at the end of basic block 1. Those events in phase B, where the issue rate is high, always cause emergencies. Those same events in phase A never do. Therefore, tracking program flow and microarchitectural events yields a proxy for the

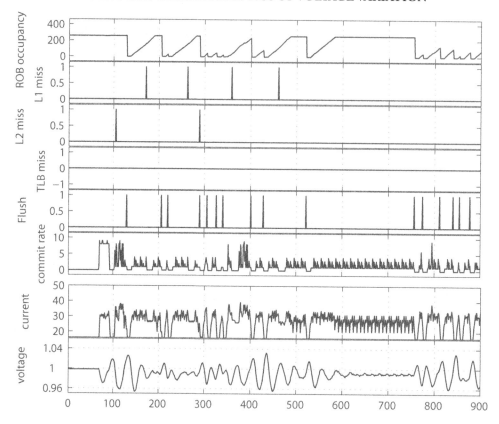

Figure 3.13: Uncharacterizable Activity. This microarchitectural activity snapshot corresponds to the `longest_match` procedure in *gzip*. No one particular event, or set of events, is the cause for the voltage variation.

activity leading to emergencies. The findings are similar for events in phase C that correspond to events in phase A.

Gupta et al. [38] devised an algorithm to automatically identify root causes in single-threaded processors. The algorithm scans recent processor events in a fixed-priority order, looking for event completion times that coincide with the time of the emergency. It scans down the list of L2 misses, TLB misses, pipeline flushes, L1 misses, and long latency operations, in that order. To show the strength of the relationship between these processor events/operations and emergencies, Figure 3.15 shows the percentage of emergencies caused by different root causes for the SPEC CPU2000 benchmark suite. A majority of the emergencies are caused by pipeline flushes and L2 misses. Interference of these events is also important. The uncharacterized 14% corresponds to emergencies that cannot be uniquely identified (i.e. mixed or others).

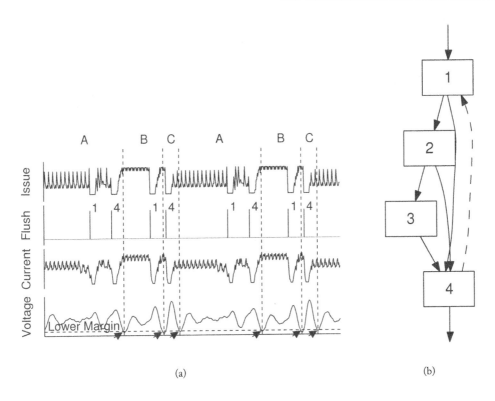

(a)   (b)

Figure 3.14: Mixed Events. (a) Voltage variation is associated with recurring activity (phases A, B, and C) over 880 cycles. The numbers next to the vertical bars in the flush graph correspond to the basic block number in (b) containing the mispredicted branch. (b) A sensitive loop from function init_regs in *gcc* from the SPEC CPU2006 benchmark suite. Its activity snapshot is shown in (a).

Interference   In multicore systems, microarchitectural activity can interfere across cores and lead to either constructive (bad) or destructive (good) interference. Researchers demonstrated this behavior on a Intel© Core™ 2 Duo processor [80] by simultaneously running microbenchmarks on each processor core and measuring the magnitude of the peak-to-peak voltage swing across the entire chip.

Figure 3.16 shows the interference heat map. Both cores are subject to different microarchitectural activity, including L1 cache misses (only) and L2 cache misses, translation lookaside buffer (TLB) misses, and branch mispredictions (BR). The magnitude of the chip-wide voltage swings are normalized relative to an idling machine. The *y*-axis corresponds to microarchitectural activity on Core 0, and the *x*-axis corresponds to activity on Core 1. The data indicates that depending on the pair of events occurring simultaneously, the magnitude of the voltage swing can vary substantially.

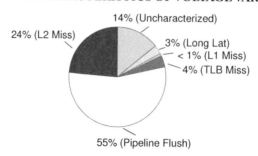

Figure 3.15: Pie Chart Distribution of Voltage Variaton Causes. Distribution of microarchitectural events and operations that cause the most voltage emergencies in the SPEC CPU2000 benchmark suite [36].

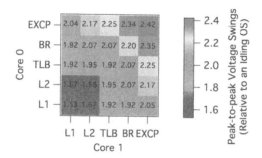

Figure 3.16: Interference. Microarchitectural event interactions in a multicore system and the resulting effect on core voltage [80], e.g., BR-BR is worse than L1-L1 pairing.

3.4 PROGRAM BEHAVIOR

Strong voltage variation can also be caused due to pure program behavior. Hazelwood and Brooks [40] showed that voltage emergencies are correlated with an application's dynamic code stream and not just the underlying architecture and power-delivery subsystem, and as such, the variation problem can also be dealt with at the application and software layer.

Problematic Loops Joseph et al.[50] were one of the first to identify and demonstrate that the most problematic processor current profiles include successive periods of high and low processor activity. It is when these high and low durations approach the resonant frequency of the power-supply network that the problem becomes more serious. The problem was demonstrated by developing an artificial application that was hand-tuned to simulate periods of high and low activity that matched the resonant frequency of the processor's power-supply network. The synthetic program shown in Figure 3.17(a) contains a single loop body that consistently causes voltage swings dangerously large enough to violate the minimum voltage margin, as shown in Figure 3.17(b).

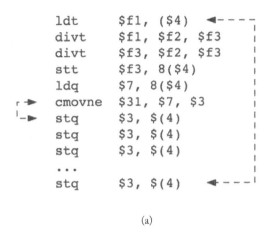

```
        ldt       $f1, ($4)
        divt      $f1, $f2, $f3
        divt      $f3, $f2, $f3
        stt       $f3, 8($4)
        ldq       $7, 8($4)
        cmovne    $31, $7, $3
        stq       $3, $(4)
        stq       $3, $(4)
        stq       $3, $(4)
        ...
        stq       $3, $(4)
```

(a)

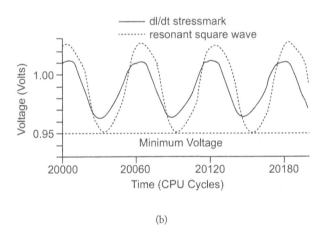

(b)

Figure 3.17: Problematic Loop Behavior. (a) $\frac{di}{dt}$ stressmark. (b) Voltage swing of the $\frac{di}{dt}$ stressmark vs. peak swings at resonance [50].

The loop body oscillates between very low current activity (because divt produces long stalls) and high current activity (in which dependent instructions store the divt result to memory, reread it, and restore it to registers). This software code loop provides motivation for software-based solutions to mitigate voltage variation, because, if such loops exist in real applications, then it is logical to apply a permanent solution at the code level, thereby limiting the performance penalty of (repeatedly) activating control hardware.

In fact, Gupta et al. [38] showed that loops are responsible for nearly all the large voltage emergencies observed within an experimental framework. Table 3.1 summarizes statistics about

Table 3.1: Problematic Loops. Loops associated with emergencies in the SPEC CPU 2000 programs

Benchmark	Total Loops	Total Emergencies	Total Emergency Loops	Procedural Emergencies
applu	479	389897	13	0
apsi	718	21056	25	537
art	293	120885	13	0
bzip	383	403798	35	6181
crafty	1406	623977	302	190181
equake	423	174293	9	6942
gap	1806	243259	48	2115
gzip	310	67324	40	17
mcf	338	88828	23	620
mesa	536	279904	41	41
mgrid	411	1111152	32	75
swim	425	2332	4	0
twolf	1271	514970	81	138241
wupwise	425	42598	14	81

the static loops present in the SPEC CPU 2000 programs. Loops that are active at the time of emergencies are the loops that we are interested in examining. The third column of Table 3.1 tabulates the number of emergency loops identified, i.e., the number of static loops in which a voltage emergency occurs. Though the total number of loops ranges from 310 to 1806, the total number of emergency loops for each application is a small fraction of this total. Most programs have a small number of procedural emergencies and more than 90% of the emergencies can be associated with unique loops. Benchmarks *crafty* and *twolf* are the only two that do not follow the trend, with 33% and 25% of the voltage emergencies classified as procedural emergencies, respectively.

Emergency Hotspots The next step toward understanding the relationship between application runtime behavior and voltage emergencies is to uniquely identify the "signature" of the emergencies, to understand whether the same signature of emergencies occurs repeatedly, and determine the frequency at which they occur. This code-region identification is required to provide the compiler

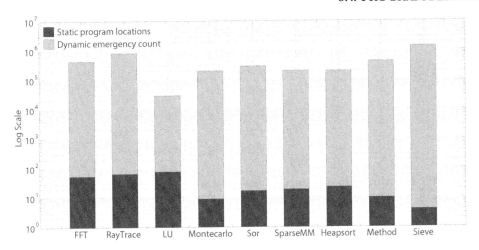

Figure 3.18: Emergency Hotspots. A small set of static program addresses (fewer than 100) are responsible for the large number of voltage emergencies. We assume a 4% operating margin, but this trend remains across different margins.

with hints in order to eliminate the problematic behavior. Hazelwood and Brooks [40] identified the use of the last executed branch (LEB) as a starting point for software to optimize the candidate code region.

Using a technique very similar to LEB that was later refined by Gupta et al. [38], Reddi et al. [75] characterized the number of distinct voltage emergencies that occurred during program execution in the JavaGrande benchmarking suite. Figure 3.18 shows the number of distinct static program locations that were deemed as voltage-emergency hotspots. The interesting finding corresponds to the number of times that the software would need to intervene to correct voltage emergencies, which is (ideally) once per static program location, vs. the total dynamic emergency count, which is the number of times that the hardware must intervene if the software is not part of the solution.

Interthread Interference   As the number of cores increases, and cores share the same power-delivery network, increasingly, one core can either constructively or destructively interfere with activity from other cores. Constructive interference is bad because it amplifies voltage variation while destructive interference is good because it dampens the variation.

Figure 3.19 is an example of the interthread voltage variation interference. The measured results are based on an Intel Core 2 Duo processor. Both cores in the experiment share the power delivery network. For such a system, the figure quantifies the aggregate droop activity where both cores are simultaneously running two instances of the *473.astar* program from the SPEC CPU2006 benchmark suite. The *x*-axis corresponds to the start time offset between the two programs. In other words, the graph is a convolution of two executions of the same program, offset by different start

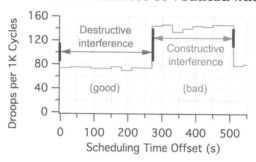

Figure 3.19: Interthread Interference. In multicore systems we observe voltage variation interference when multiple threads run simultaneously [80].

times. Voltage variation, expressed as droops per 1K cycles, when two cores are active, is smaller between 0s and 275s than later (between 275s and 500s). The former is an example of destructive interference. The latter is an example of constructive interference. Interfering microarchitectural activity across cores, such as pipeline flushes and cache misses, is the root cause of both the constructive and destructive interference. We discussed microarchitectural event interference previously in Section 3.3. Similar behavior was observed across other programs in the SPEC CPU2006 suite [80].

At first glance, constructive thread alignment would seem to be a low-probability event in multi-core machines with complex, out-of-order cores and shared and non-shared resources. However, Kim et al. [54] showed that alignment occurs relatively often when threads consists of short execution loops. The authors observed this behavior during droop measurements of an AMD processor and determined that the effects were caused due to natural perturbations in the threads as a result of the operating system's thread scheduling dynamics.

Kim et al. showed an example of naturally occurring behavior over the course of 100 ms when running a four-threaded loop-intensive program that exhibits resonant behavior within each loop. Each major grid point in Figure 3.20 corresponds to 10 ms and the y-axis shows the measured processor voltage (V_{DD}) values at a 100 MS/s sampling rate. Approximately every 16 ms V_{DD} variability changes, which corresponds to the OS timer tick on the Windows operating system. When the threads align constructively, as is the case around the center point of the scope shot, the voltage droop is maximized. This data shows that repetitive loops across multiple threads at the same time can result in significant voltage variation in the system, due to naturally occurring interthread execution interference.

3.5   SUMMARY

As the industry trends toward aggressive power management and voltage scaling in future multicore designs, it is increasingly important for architects to understand the potential for voltage variation. To this end, we took an in-depth look at factors that influence the magnitude of voltage variation

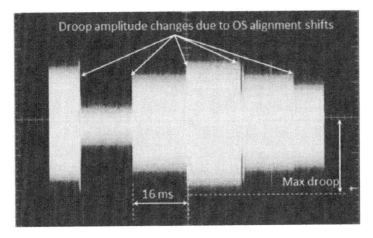

Figure 3.20: Thread Alignment Effects. The magnitude of the variation increases or decreases depending upon voltage plane interactions across threads [54].

in a system. These factors range from current pulse amplitude and periods to the characteristics of the power-delivery network combined with microarchitectural processor activity and program behavior. In effect, voltage variation is the result of complex interactions between the different factors. Therefore, while we explored the individual factors' impact by looking at specific examples, it is important to understand that the exact behavior can vary depending on the complex interactions between all factors.

CHAPTER 4

Traditional Solutions and Emerging Solution Forecast

There are two fundamental approaches to dealing with voltage variation. Broadly speaking, there are static techniques and dynamic techniques. In this chapter, we first give a brief overview of the traditional approaches that specifically deal with voltage variation (i.e., voltage margins, floorplanning, and adding decoupling capacitance at various levels of the system). However, these are static techniques that do not take into account dynamic activity that impacts voltage variation. In the previous chapters, we discovered that voltage variation is the result of complex interactions between PDN characteristics, application activity, and microarchitectural event behavior. Recently, several dynamic approaches have been proposed that leverage knowledge about these interactions. We grouped these approaches into the following categories: tolerance, avoidance, and elimination. Before elaborating on the various techniques, in this chapter we provide the key insight behind each of those categorical approaches. We provide the intuition behind each of these approaches, and subsequently elaborate on the techniques in the following chapters.

4.1 TRADITIONAL STATIC TECHNIQUES

Designers must take several precautions to ensure voltage variation is kept to a minimum during operation. Current designs prevent dangerous voltage fluctuations via careful allocation of large voltage margins, placement of decoupling capacitors, and advanced floorplanning. These are all static techniques that designers have applied traditionally, and which we must adapt.

4.1.1 VOLTAGE MARGINS

Today's production processors use operating voltage margins that are nearly 20% of nominal supply voltage [49]. However, conservative designs either lower the operating frequency or sacrifice power efficiency. As feature sizes shrink and nominal supply voltage scales down gradually with limited threshold voltage scaling, circuit delay sensitivity to margins increases with each technology node.

Figure 4.1 plots peak frequency at different voltage margins across four PTM [99] technology nodes (45 nm, 32 nm, 22 nm, and 16 nm) based on detailed circuit-level simulations of an 11-stage ring oscillator consisting of fanout-of-4 inverters. The plot shows that at today's 32 nm node, a 20% voltage margin translates to a 33% frequency degradation, and at future technology nodes, the

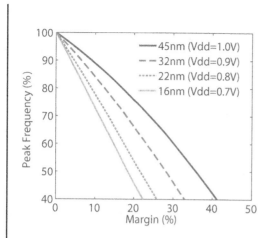

Figure 4.1: Impact of Worst-case Margins. Worst-case voltage margins are a growing source of processor inefficiency [79].

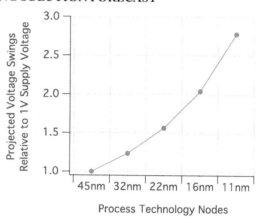

Figure 4.2: Impact of Technology Scaling. Voltage variation is a growing problem in future process generations, as the peak-to-peak swings are increasing [80].

situation gets much worse. Practical limitations on reducing power-delivery impedance, combined with large current fluctuations, make margin-based solutions unsustainable.

Trends indicate that margins will need to grow to accommodate worsening peak-to-peak voltage swings. Consequently, designers must increasingly compromise peak performance for growing worst-case delays. Figure 4.2 shows the worst-case peak-to-peak swing in future generations relative to today's 45 nm process technology. This data is based on simulations of a Pentium 4 power-delivery package model [37], assuming nominal voltage gradually scales according to ITRS projections from 1 V in 45 nm to 0.6 V in 11 nm [48]. To study package response, current stimulus goes from 50 A to 100 A in 45 nm. Subsequent stimuli in newer generations is inversely proportional to V_{DD} at the same power budget. Voltage swing doubles by the 16 nm technology node. Future processor performance and power efficiency will suffer to an even greater extent than in today's systems.

4.1.2   DECOUPLING CAPACITORS

To dampen voltage variation and keep voltage margins within some reasonable bounds, processor designers also rely on package and on-chip decoupling capacitance. These capacitors attempt to maintain low impedance over a range of frequencies. Bulk capacitors on the motherboard dampen low-frequency noise, while package capacitors target mid-frequency variation between 50 and 200 MHz that is caused by impedance in the power-delivery network. Lastly, on-chip decoupling capacitance targets high-frequency variation caused by sudden sharp changes in current due to dynamic clock gating of idle functional units. Figure 4.3 illustrates the distribution of these capacitors over the different types of voltage droops.

For today's processors, designers need to apply a significant amount of on-chip decouple capacitance to keep the magnitude of voltage swings within tolerable bounds. However, there is a careful trade-off to be made between the amount of decoupling capacitance that is used and the area that the decoupling capacitance requires. For instance, researchers [21] demonstrated estimates that if the CMOS thick-oxide gate provides 10 $fF/\mu m^2$ [71], an area of 20 mm^2 is needed to provide 200 nF of decoupling capacitance. The Alpha 21264 processor's designers reported that roughly 15% to 20% of the die area is occupied by decaps [34]. Therefore, it is important to estimate and allocate the area needed for on-chip decoupling capacitors during the early design stage.

Although designers have traditionally used oxide capacitors [57], the industry is making advances toward integrating deep-trench decoupling capacitors [44] into logic circuits. Deep-trench capacitors provide significantly more capacitance per unit area than oxide capacitors. Figure 4.4 shows the frequency response of package impedance for thick-oxide capacitors vs. deep-trench capacitors. With the integration of deep-trench capacitors into logic circuits, it is conceivable that off-chip decoupling capacitors on the package may be completely eliminated in the future. However, deep-trench decoupling capacitance does have its drawbacks. The area requirements are high. Deep-trench capacitors will also add to a chip's cost, due to the costly manufacturing process, and will exacerbate the already problematic leakage power problem in processors.

4.1.3 FLOORPLANNING

Modules within the processor do not exert uniform current demands. Some modules consume significantly more power than others. It is critical to ensure that such modules have a low impedance path on the power-delivery grid. Similarly, it is also important that designers do not place high-power modules that are likely to simultaneously switch on or off close together. Such placement could lead to a sudden large current swing in a short amount of time, causing a voltage emergency. Therefore, it is possible to engineer a floorplan that is resistant to such voltage variation by distributing the current demand of modules more regularly across the processor [22, 64, 70]. Because modules within the processor do not have uniform current demand, designers can exploit this information to place high-current modules spatially far apart from one another by pairing them with low-power modules.

Overall, margins, decoupling capacitance, and floorplanning all help make the processor robust against voltage noise. However, such static solutions require careful preplanning. For instance, at present, the only quantifiable methodology that strongly establishes the amount of decoupling capacitance required involves planning for the worst-case voltage swing. Such pessimistic design lowers the processor's overall operating efficiency. The traditional means of dealing with voltage variation discussed in this section are already being stretched to their limits. Continued scaling trends will only make voltage variation a more serious problem for the community to address. As performance, power efficiency, area, and cost become more important, new and more cost-effective solutions will become necessary to cope with all forms of variations.

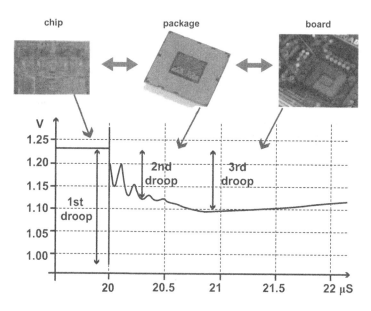

K. Wong et al., JSSC, 2006

Figure 4.3: Voltage Droop Types and Capacitances. Sources of capacitance for the three primary types of voltage droops.

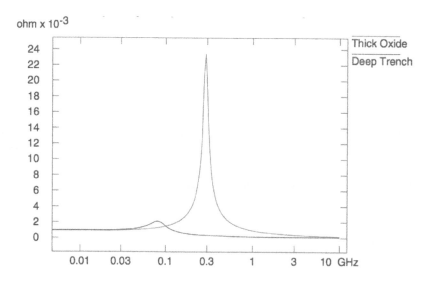

Figure 4.4: Impact of Capacitance Type. Effect of deep-trench decoupling capacitance on package impedance.

4.2   TOWARD DYNAMIC TECHNIQUES

While current designs are able to manage voltage variation through careful placement of decoupling capacitors and advanced packaging, traditional means of reducing voltage variation are being severely stretched due to recent technology trends. These trends, that will make voltage variation management a considerable challenge in the coming few years, include increasing processor currents, decreasing supply voltages, and a significant increase in current variability due to power-saving gating techniques [46].

The problem with addressing voltage variation using the static techniques discussed in the previous section is that such techniques are overly conservative. The solutions put in place are not adaptable after the chip hits the market. Therefore, designers make cautious and pessimistic assumptions about the conditions under which a chip may operate to ensure high reliability. But such conservative design strategies lead to worst-case design that is nonrepresentative of typical-case behavior, causing poor performance and power inefficiency.

Prior work demonstrated that worst-case design is overly conservative for a production processor from 881 benchmarking runs on an Intel© Core^{TM} 2 Duo [80]. Figure 4.5(a) shows a cumulative histogram distribution of voltage samples for the Core 2 Duo processor. The figure shows the deviation of each voltage sample relative to the nominal supply voltage. Each line within the graph corresponds to a run. The 881 runs include a spectrum of workload characteristics: 29 single-threaded SPEC CPU2006 workloads, 11 Parsec [12] programs, and 29×29 multiprogram workload combinations from CPU2006.

Runtime voltage variation for these programs is as large as 9.6% (see min. droop marker), and therefore the estimated 14% worst-case margin is necessary for this processor. But such large variations occur infrequently. Most voltage samples are within 4% of the nominal V_{DD}. The typical-case marker in Figure 4.5(a) indicates this range. Only a small fraction of samples (0.06%) lie beyond this region. Therefore, it is a better design choice to tighten the worst-case voltage margin to 4%, while providing a fail-safe guarantee mechanism for those very infrequent large voltage swings.

In the following sections, we will introduce the fail-safe mechanisms under the umbrella of the following three categories: tolerance, avoidance, and elimination. Here, our objective is to first explain the driving philosophy behind each of the categorical approaches. In the chapters that follow, we will present the myriad of techniques under each category.

4.2.1   TOLERANCE

Tolerating or allowing voltage emergencies to occur, rather than preventing them at any cost, is useful for two important reasons. First, it enables designers to tighten the voltage margin. Second, it enables feedback-driven solutions to the problem, i.e., by observing the emergency behavior of running code the system can learn to adapt itself to avoid emergencies. In order to enable tolerance, the architecture must support a built-in mechanism that allows voltage emergencies to occur, but, when they do, it must recover state and resume.

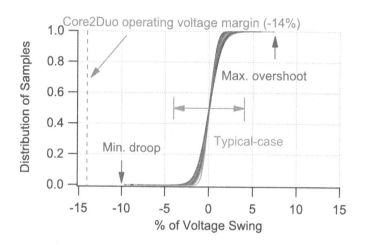

(a) Worst-case vs. typical-case voltage variation.

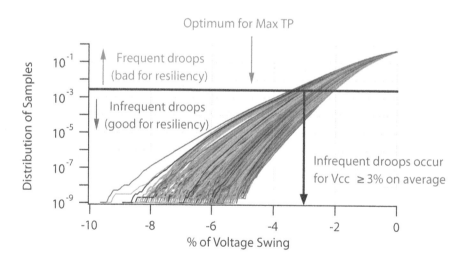

(b) Wide tail distribution between programs.

Figure 4.5: Measured Voltage Variation Behavior in a Typical System. Measured cumulative distribution of supply voltage (V_{DD}) samples vs. the percentage of nominal V_{DD} for 881 unique programs executing on the Intel Core 2 Duo processor [80].

Some researchers have proposed dedicated recovery schemes to tolerate *all* voltage emergencies [39]. Such schemes are typically intrusive and require changes to traditional microarchitectural structures, complicating design and testing, in addition to requiring revalidation of prior logic blocks. However, they are extremely efficient at tolerating voltage variation. In contrast, more general-purpose coarse-grained checkpoint-recovery units are already available in existing production systems [3, 86]. These systems have been leveraged to serve numerous other purposes, such as testing and debugging [55, 61, 67, 85, 89, 96]. Therefore, an alternative to using custom recovery blocks is to rely on these general units.

While we may think that general-purpose checkpoint recovery is good from a reusability standpoint, the cost of relying purely on coarse-grained checkpoint-recovery is prohibitively expensive. Tolerating emergencies is not always possible. The intervals of traditional checkpoint-recovery schemes (between 100 and 1000 cycles) translate to unacceptable performance penalties. Therefore, it is only feasible to rely on this mechanism infrequently.

4.2.2   AVOIDANCE

Tolerating emergencies using coarse-grained checkpoint-recovery hardware is not always practical, because it can be a prohibitively expensive rollback mechanism. Therefore, researchers have proposed a variety of emergency predictors that identify when emergencies are imminent and prevent their occurrence by throttling execution. Throttling is the act of slowing down machine execution so that voltage recovers to its nominal level gracefully. A voltage-emergency predictor can predict voltage emergencies using a variety of heuristics, such as current and voltage profiles, microarchitectural activity signatures, and so forth.

An added benefit of building avoidance mechanisms is that it is possible to train the microarchitecture. The processor can learn from emergency activity occurring during runtime and tweak its behavior (e.g., issue rate) in response to that behavior. In the previous chapter, we discussed several microarchitectural events and code characteristics that can induce strong voltage variation. Such knowledge, applied systematically, can enable strong prediction mechanisms that can anticipate emergencies based on little information.

Avoidance mechanisms can be a form of feedback-driven optimization that is beneficial because applications exhibit different voltage variation behavior. Figure 4.5(b) represents a zoomed-in version of Figure 4.5(a). The figure demonstrates that the distribution of V_{CC} droops varies widely across the 881 programs. V_{CC} droop magnitudes that occur less than 0.25% ($V_{CC} > 3\%$) are infrequent and thus, can be effectively mitigated by an avoidance mechanism. However, V_{CC} droop magnitudes that occur greater than 0.25% ($V_{CC} < 3\%$) are too frequent to reduce with any resilient hardware. But, this data indicates there is there is room for runtime specialization based on application behavior.

4.2.3   ELIMINATION

Better than tolerating or avoiding emergencies is eliminating them altogether. Consider a frequently executing loop that experiences recurring emergencies every iteration of the loop because the program is taking the same error-prone code path every iteration. Such a scenario can be handled better in software than in the hardware. Hardware would need to repeatedly tolerate or throttle processor execution in order to avoid the emergency. But an intelligent piece of software, such as a compiler, can perform fine-grained instruction-level code transformations to eliminate the emergency. A compiler typically has several options when choosing the order of instructions and many of the options result in equally performing software. Therefore, in the case of a voltage-emergency-prone loop, such as Joseph et al.'s code in Figure 3.17(b), the compiler can rearrange instructions along the problematic code path to avoid recurring emergency activity without impacting performance.

Figure 4.6 shows the contribution to total emergencies from loops. Almost all the voltage emergencies occur in loops. This suggests that the runtime overhead of voltage-specific software optimizations would be small, because the overhead can be easily amortized. We optimize the loop once and reap the rewards of emergency-free loop execution for the remainder of the loop's iterations. In this context, we are also interested in the loops that incur the greatest numbers of voltage emergencies. Therefore, in Figure 4.6 we show the percentage of total voltage emergencies that occur in the top five emergency loops for each SPEC CPU 2000 program. For each program, around 2 and 5 loops account for more than 75% of the emergencies and, hence, we can classify these loops as hot loops.

The data above indicates that optimizing these hot loops can significantly reduce the overall number of emergencies. Identifying such problematic hot loops, understanding the characteristics of these loops, and investigating their interaction with other loops can provide insights into the sequences of events that lead to emergencies. This understanding is critical to finding perturbations to the code that will eliminate the emergencies.

In a multicore processor, it may be difficult to coordinate and orchestrate individual core activity in hardware to anticipate emergencies. For instance, in Section 3.1 we examine how local and global resonant current pulses interact. One key observation was that, even when there is no locally resonant core activity, there can be global resonance. As the number of cores increases, hardware coordination is likely to be problematic. However, a firmware-level thread scheduler can smooth out voltage variation across interfering execution threads.

Thread scheduling is an important topic of study in symmetric or chip multiprocessors (CMP). Prior work demonstrated that threads can hurt each other's performance by destructively interfering with one another [19, 20, 31, 56, 60, 88, 100]. For instance, scheduling two cache-intensive programs together is a bad decision, because the shared cache resource becomes a bottleneck and both programs suffer. And as such, it is better to schedule one of those cache-intensive programs with another program that is more CPU-bound (i.e., less intensive on the cache), resulting in less interference and better overall system performance.

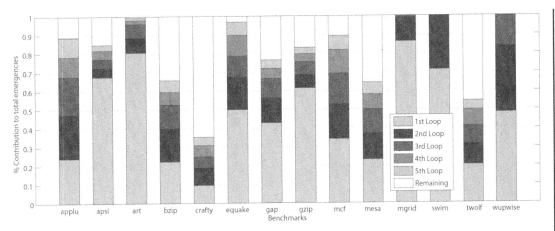

Figure 4.6: Emergency-prone Loops. Contribution of loops (top 5) to voltage emergencies in SPEC CPU2000 programs.

In the previous chapter, we observed that similar interthread interference exists in the context of voltage variation. Therefore, a variation-aware thread scheduler can schedule threads intelligently to minimize emergencies. By reducing emergencies, a system's overall throughput increases due to fewer rollbacks; in our system, we assume a global checkpoint-recovery mechanism across all cores sharing a power supply source.

4.3   SUMMARY

In the future, we will require better integration and collaboration between devices, circuits, architecture, and the software layer, building on the categorical principles of tolerance, avoidance, and elimination. As technology trends force us toward typical-case designs, error-detection and recovery mechanisms will both need to become pervasive in microprocessor designs. Sustained increases in performance and energy efficiency require us to identify and develop new dynamic techniques that can dynamically detect and recover from errors in the field to overcome the penalties of traditional static techniques.

CHAPTER 5

Allowing and Tolerating Voltage Emergencies

In this chapter, we explore proposed architectural mechanisms that enable the performance and power the benefits of typical-case design margins, specifically, by tolerating voltage emergencies. Rather than trying to prevent voltage emergencies with the aggressive use of static techniques, the concept of tolerance allows margin violations to occur, but, when they do, the architecture has the ability to roll back to a guaranteed-correct processor state.

There are two options for implementing tolerance in today's microprocessors. The first option is to leverage traditional checkpoint-recovery schemes built for recovering from soft errors. The second option is to develop a specialized recovery solution that is specifically designed and targeted toward tolerating voltage emergencies. We will look at a system called the delayed commit and rollback mechanism (DeCoR) to handle voltage emergencies.

5.1 ERROR DETECTION

A key requirement of any tolerating mechanism is timely detection of timing violations, or voltage emergencies, in order to prevent propagation of corrupted state across unit boundaries. To maintain correct semantics and ensure that a corrupt instruction does not propagate from the unit in violation of the rest of the pipeline, errors must be detected at the transition boundaries between units. Here, we present two basic circuit designs for timing-error detection. First, we discuss an embedded error-detection sequential (EDS) circuit [16, 17, 23, 24, 30, 32, 33, 68]. Second, we discuss a tunable replica circuit (TRC) [17].

Error-Detection Sequential Circuit In the embedded EDS design, an EDS circuit (shown in Figure 5.1(a)) replaces the receiving flip-flop of critical paths to detect late timing transitions during the high-clock phase. During normal operation (i.e., without variations), the input data typically arrives well before a preset setup-time constraint. However, in the presence of dynamic variations, the input data may arrive slower at the receiving flip-flop. Note that the input data to the receiving flip-flop must always arrive setup-time prior to the rising-clock edge in order to ensure correct functionality. In order to detect slow- or late-data arrival, an EDS circuit double samples the input data with a datapath flip-flop on the rising-clock edge and a shadow latch on the falling-clock edge. The flip-flop and latch outputs are compared with an XOR gate. In the event of a mismatch, an error signal is raised to the the microarchitecture level to enable error recovery.

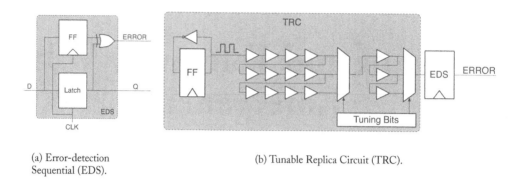

(a) Error-detection
Sequential (EDS).

(b) Tunable Replica Circuit (TRC).

Figure 5.1: Timing Error Detection Schemes. Both are useful for detecting emergencies.

The fundamental trade-off in the EDS circuit is max-delay vs. min-delay. The width of the error-detection window determines the maximum potential benefit in F_{CLK} as well as the min-delay penalty. For a target error-detection window, presilicon design satisfies the min-delay requirements with appropriate buffer insertion and sizing. In addition, a scan-configurable duty-cycle control circuit allows designers to perform post-silicon tuning of the high-clock phase to maximize the error-detection window while avoiding min-delay errors. During runtime, an OR tree combines the error signals from each EDS circuit in a pipeline stage to produce a single pipeline-error signal, which can be pipelined to the write-back stage in order to invalidate an errant instruction and to initiate error recovery [17].

Tunable Replica Circuit   In comparison to the embedded EDS design, the TRC design (shown in Figure 5.1(b)) is less intrusive [17]. The TRC detects timing failures for a scan-configurable buffer delay chain with the input transitioning every cycle. A TRC is placed adjacent to each pipeline stage. Post-silicon calibration of the buffer delay chain ensures that the TRC always fails if any critical path fails in the pipeline stage due to a dynamic variation. Thus, the TRC is tuned slower than the critical path.

The TRC and the paths in the pipeline stage use the same local V_{CC} and clock, which enables the TRC to track critical-path delay changes due to V_{CC} droops while capturing clock-to-data correlations. If a dynamic variation induces a late timing transition in the TRC, the circuit generates an error signal that represents the single pipeline-error signal. As with the embedded EDS design, the single pipeline-error signal is pipelined to the write-back stage to invalidate the errant instruction and to enable recovery.

EDS vs. TRC   Table 5.1 lists the key trade-offs between the embedded EDS and TRC designs. The EDS design detects critical-path timing failures for fast, slow, long-range, and local dynamic variations. In contrast, the TRC design cannot detect path-specific or highly-localized dynamic vari-

Table 5.1: Comparion of Schemes. EDS vs. TRC design trade-offs [17]

	EDS Design	**TRC Design**
Dynamic Variation Detected	Slow & fast, long-range & local	slow & fast, long-range
Exploit Path Activation	Yes	No
False Error Recovery	No	Yes
Design Complexity	High	Low
Min-Delay Overhead	Yes	No
Clock Energy Overhead	Yes	Negligible
Clock Duty-Cycle Control	Required	Not required
Maximum Potential Benefit	Limited by core min-delay paths	Not limited by core min-delay paths
Post-Si Calibration	Clock duty-cycle	TRC delays

ations (e.g., delay push-out from cross-coupling capacitance). The TRC requires a delay guardband to ensure the TRC delay is always slower than critical-path delays, thus preventing the possibility of exploiting path-activation rates for higher performance. Furthermore, the TRC design may initiate an error recovery when an actual error did not occur. This results in unnecessary recovery cycles. In comparison to the EDS design, the TRC design significantly reduces the design complexity overhead. In particular, the TRC design does not affect the min-delay paths in the core, has lower clocking energy, and does not require a duty-cycle control circuit. While the core min-delay constraints limit the error-detection window for the EDS design and, consequently, the maximum potential benefits, the TRC design provides a larger error-detection window to detect a wider range of dynamic delay variation. Both designs require post-silicon calibration, which affects testing.

5.2   GLOBAL RECOVERY

Existing processors have begun to implement global recovery (GR) mechanisms to handle errors, such as the recovery unit (RU) in POWER6 [59]. There can be several mechanisms for explicit checkpoints that vary in their degree of implementation complexity and overheads involved. Earlier checkpointing schemes were predominantly aimed at providing fault tolerance in large-scale, high-performance computing systems. Recently, however, checkpointing schemes have been proposed for several other domains: soft-error detection [96], boosting processor performance [55, 61, 89], fault detection [85], and debugging [67]. In this section, we will discuss efforts to expand checkpoint recovery to tolerating voltage variation.

5.2.1   CHECKPOINT RECOVERY

A common trait of such checkpoint-recovery schemes is the explicit saving and restoring of required architectural state. The state that must be saved for correct execution at recovery is mainly the architected state, which consists of the registers and the updated memory state. For example, [55] and [61] assume a buffered memory update, using volatile bits for updated lines between checkpoints. When a rollback occurs, the lines marked volatile are flushed from the cache. Of course, this results in additional cache misses after rollback, but, at least resumed execution remains correct. The overhead of implementations proposed in [1] and [61] includes a register restore latency of eight cycles (for 32 registers with four write ports). The infrequency of the rollbacks in such implementations typically shadows the cost of supporting a rollback scheme. These schemes aim to take checkpoints at very coarse granularities, ranging between 100 to a few 1000s of processor-clock cycles.

General-purpose checkpoint recovery is good when rollbacks occur infrequently, such as in the case of soft errors. Applied to voltage emergencies, the cost of frequent recovery can outweigh the benefits of using tighter voltage margins for typical-case design. Soft errors and voltage emergencies are similar in that they can cause transient failures, but they differ greatly in their characteristics. The main differences are in (1) the physical phenomenon that causes them, (2) the frequency of error occurrence, (3) the structures sensitive to the errors, and (4) the relationship between application characteristics and error occurrence. Understanding these differences and their requirements for detection and correction is important for knowing whether GR is applicable for voltage emergencies.

Soft errors are transient errors that are generally characterized as single-event upsets or bit flips, caused by energetic particles from cosmic rays or alpha particles. The occurrence of soft errors is quite rare, and these errors primarily affect data-storage nodes. The probability of single-event upsets affecting the correctness of computation depends on the architectural vulnerability factor of the logic [65], which determines whether a fault in that logic would actually affect the application's outcome. A common strategy is to employ a reactive mechanism where appropriate recovery actions are taken once an error has been detected. A main challenge in dealing with soft errors is the implementation of a robust error-detection mechanism, often seen in the form of parity bits and/or error-correcting codes (ECC). For example, a parity-bit propagation technique to detect soft errors was implemented by Fujitsu in their SPARC processors, providing coverage for almost 80% of latch banks and array structures [4].

Other detection approaches use redundant (or checker) processors and threads, which re-execute some, or all, of the instructions to verify correctness [6, 82, 90]. The infrequent occurrence

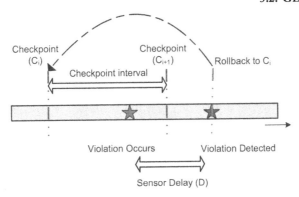

Figure 5.2: Tolerating Sensor Delay Via Two Checkpoints. Example illustrating that two checkpoints are required in explicit checkpoint schemes and are used to maintain correctness in the presence of voltage emergencies.

of soft errors allows these reactive mechanisms to have large penalties associated with recovery. For example, the Fujitsu processor employs a checkpoint hardware mechanism with a quiescent and preparation period for restart of around 1μs. Such microsecond-scale penalties are acceptable for soft errors that occur at the timescale of days.

Voltage emergencies are also transient errors, but they have different characteristics from those observed for soft errors. Voltage emergencies are dependent upon the interactions across, and the characteristics of, the underlying power-delivery network, the processor's microarchitecture, and the executing application. Recall that voltage variation results from parasitics present in the system that can cause voltage swings in response to current fluctuations. If the voltage swings are significant, they can induce timing-margin violations. Voltage emergencies primarily affect logic delay paths and are tightly coupled to application characteristics. For example, the presence of repeated execution patterns in applications can increase the susceptibility to timing violations due to resonance in the power-delivery network [40]. Unlike soft errors, noise-margin violations are easier to detect (e.g., using hardware sensors), but require careful balance between latency and resolution.

A subtle, but important, aspect of handling voltage emergencies using general-purpose checkpoint recovery is that the recovery scheme must be invariant to sensor delays. Detecting emergencies is not instantaneous. Voltage data from a number of sensors must be continuously gathered and analyzed to detect a margin violation. This can be in the order of five to seven clock cycles, or more, which requires the processor to maintain two checkpoints for correctness. Figure 5.2 shows an example of a scenario where the voltage emergency is detected after checkpoint C_{i+1} has been taken. To have correct semantics, the system should rollback to checkpoint C_i, which incurs more rollback penalty because it is further back in time. In general, even if the checkpoint interval is adjusted to match the detection delay in the checkpoint-recovery mechanisms, the recovery costs of discrete, explicit mechanisms are higher, and thus require more implicit checkpointing techniques.

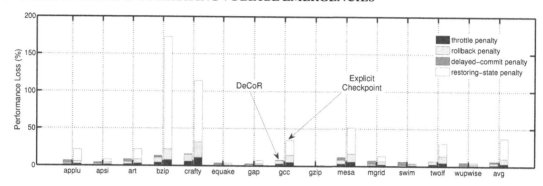

Figure 5.3: Penalty Analysis. Breakdown of different penalties associated with our proposed delayed-commit and rollback scheme and explicit checkpoint schemes. The sensor delay assumed for this comparison is 20 cycles and checkpoint interval of 21 cycles for explicit checkpoint schemes.

Figure 5.3 shows a breakdown of the performance impact of rollback, throttle, and restoring of state for explicit checkpoint and recovery for a conventional out-of-order superscalar x86 processor with a Pentium 4 package [7], assuming a typical-case 5% voltage margin. Note that the system must throttle the processor after a rollback in order to guarantee that the same burst of activity that first caused the emergency does not reoccur. From the data we observe that, due to the rollback to the previous checkpoint, C_i, instead of the current checkpoint, C_{i+1}, for every rollback, the rollback penalties (including the throttle penalties) are higher for the explicit checkpoint schemes. We can also see that the performance impact of restoring state in explicit checkpoint mechanisms is significantly higher, around 39% on average. This includes both the register-restore penalty as well as the impact of flushing the volatile lines. For example, a huge performance loss of 170% is seen for *bzip*, because frequent flushing of volatile lines significantly increases cache-miss rate. All programs clearly do not favor explicit checkpointing due to the large performance degradation. Clearly, explicit checkpoint mechanisms incur unacceptable performance overheads when applied to highly frequent transient errors that are characteristic of voltage variation.

5.2.2   DELAYED COMMIT AND ROLLBACK

Given the poor performance of coarse-grained checkpoint-recovery for voltage emergencies, Gupta et al. [39] proposed a more specialized recovery mechanism to tolerate emergencies. The mechanism, delayed commit and rollback (DeCoR), does not attempt to avoid voltage emergencies, as is expected of a tolerating mechanism. It lets the processor core run freely with more aggressively set timing margins and provides safeguards to detect and recover from potential timing-margin violations, if and when voltage emergencies do occur.

Protection Zones   In DeCoR, the overall machine architecture is divided into two protection zones: a zone that is timing-margin protected (TM-protected), as is traditionally done, and a zone that is

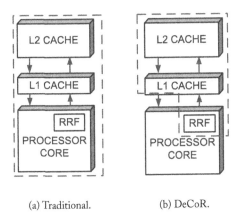

(a) Traditional.   (b) DeCoR.

Figure 5.4: Traditional Timing Protection vs. DeCor. (a) Traditional worst-case design margining requires large-enough margins to protect all processor structures inclusively. (b) With DeCoR, the required margins are for a smaller portion of the processor. The rest of the core state is recoverable using the rollback mechanism.

rollback protected (RB-protected). A RB-protect zone includes all structures where DeCoR permits recovery from voltage-induced timing violations. In particular, the vast majority of the processor core, along with the read path of the L1 data cache, is under the RB-protect zone. A TM-protected zone encompasses all other structures that require timing margins large enough to prevent voltage emergencies. The L1 cache's write path, the entire L2 cache and the retirement register file (RRF) are TM-protected.

Figure 5.4 contrasts two schemes to handle voltage emergencies. In standard design flows, the designer is responsible for meeting timing margins across the entire processor under its worst-case conditions, as shown in Figure 5.4(a). As we have seen in the previous chapters, this leads to a robust but overdesigned system for typical conditions. In DeCoR, only a relatively small part of the processor is TM-protected, thus reducing the penalties.

The TM zone relies on standard circuit-based techniques to guarantee that all timing margins are met. Although circuitry in this zone requires a more conservative design, the blocks that reside here are limited to the retirement register file, the PC chain, the L1 write port, and the L2 cache. Fortunately, these structures tend to be less timing-sensitive for several reasons. First, processor performance is relatively insensitive to L2 cache latency, so circuit-level access time is not critical to system performance. For example, many designs will construct L2 caches using low-leakage, high-V_t transistors that trade access latency for reduced power consumption [66]. For the L1 cache, the L1 *read* path is known to set the access time, while the write path is less critical [97]. This is because the read ports are driven by small SRAM cells that cannot easily be sized up, while the write ports are driven by external peripheral circuits that can be appropriately sized to increase speed.

Idle memory cells are assumed to be resilient to common-mode voltage fluctuations, which affect both sides of a differential SRAM cell equally. Intel's cache [83] shows that memory cells retain state in lower-voltage drowsy modes, where the voltages are much lower than the assumed low-end voltage-emergency level. Furthermore, idle memory cells typically have additional protection from standard ECC measures common in today's microprocessors. Finally, the retirement register file and PC chain are relatively small structures, which are unlikely to be timing-critical and can be sized up with small power penalties, if needed.

The rest of the processor pipeline resides in the RB-protected zone, which includes the instruction fetch unit, instruction cache (there are no writes to the I-cache from the processor and the static memory cells are robust, as explained above), the issue logic, the execution units, and the commit logic that consists of the reorder buffer and store queues. These structures can assume more-aggressive timing margins to avoid unnecessary performance loss, because they rely on an architectural mechanism for protection. Note that updates to the branch predictor in the speculative state may corrupt the predictor state if a rollback is initiated. However, such entries would be few, would have only slight performance impact, and will not affect correctness. Splicing the processor into TM-protected and RB-protected is also straightforward, and can simply be applied at the architectural block level (in RTL). The paths from these blocks can be flagged for extra timing margin.

Delayed Commit   To deal with voltage emergencies in the RB-protected zone, delayed-commit guarantees correctness in the presence of voltage emergencies. The delayed-commit mechanism speculatively buffers processor updates to the machine state (register file and memory) until it has verified that no emergencies have occurred during a time period sufficient for the sensors to detect an emergency. At the end of the sliding window of time, the state is said to be verified and can be committed to its respective structure.

In the event of an emergency, speculative updates are discarded and execution is restarted from a prior verified state. Thus, the delayed-commit mechanism distinguishes the processor's speculative state from its verified state. While the proposed delayed-commit and rollback mechanism may appear to resemble traditional checkpoint-recovery schemes, there are distinct differences in implementation requirements/challenges and resulting performance penalties. Figure 5.2 shows that DeCoR outperforms explicit checkpointing.

Figure 5.5 presents a functional diagram of the delayed-commit architecture. The program can be rolled back to the verified state following a timing-margin failure, which will be signaled by a voltage-sensor reading. In the speculative state, the processor continues executing the regular execution path and results are held in existing buffering mechanisms (reorder buffer (ROB) and store queue (STQ)) until the outcome of the sensor reading is known. Throughout execution, the retirement register file (RRF) and the L1 data cache hold the correct state of the program.

To know when the state becomes a verified state, each entry in the ROB/STQ has a counter associated with it. Completed results must be buffered in the ROB/STQ until they are verified to be safe. The time the instructions need to be buffered is directly proportional to the system's sensor

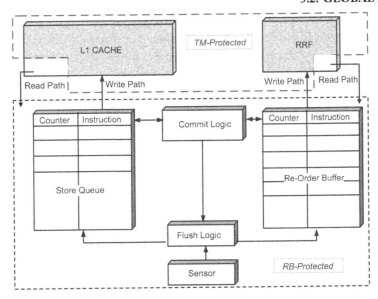

Figure 5.5: Architecture of DeCoR. A part of the RB-protected zone has been shown along with the additional modification, including the counters added to the queues. The L1 cache and register file lie in the TM-protected zone.

delay. If sensor delay is D cycles long, then all completing instructions will set their counters for this delay. When the instruction reaches the head of the queue and it is ready to retire, the commit logic verifies that the counter has expired and then declares the state as verified. At this point, it is safe to commit the state to the appropriate TM-protected structure, i.e., RRF or L1 data cache.

This scheme's correctness relies on proper transitions from the speculative state to the verified state. The transition takes place when the state is committed from the STQ to the L1 data cache and from the ROB to the RRF. The designers must guarantee the robustness of writes to the RRF and data caches at all times, because in a worst-case scenario, a voltage emergency could occur while state is being moved into the verified locations. For this reason, the designer must ensure that the write paths of these structures have sufficient timing margins to tolerate voltage emergencies, and, hence, lie in the TM-protected zone. Reads from the data cache and the retirement register file, however, can experience voltage emergencies without correctness concerns, because these emergencies would be detected and handled by the delayed-commit mechanism. Thus, the designer does not need to take any special measures when data transitions into the RB-protected zone.

Rollback   When the system detects a voltage emergency, all speculative state must be flushed. The system needs to initiate a rollback to the last verified correct state. Flushing is straightforward, as the speculative state is already located in structures (the ROB and STQ) that are capable of flushing speculative states and rolling back program execution. Thus, the rollback mechanism flush is similar to a flush after branch mispredicts, and the machine can be restarted the next cycle. A key attribute of the scheme is that rollback occurs only when emergencies actually occur; false alarms never occur, which is a key trade-off with mechanisms that we will discuss in the following chapter.

To ensure that the processor does not cause new emergencies during rollback execution, the processor starts at a reduced frequency for some number of cycles, called the throttling period. This guarantees that the program progress forward, but at the cost of some performance loss. Typically, half-rate throttling is used during rollback. This can be achieved without the PLL's involvement—that is, by gating the clock every other cycle.

5.3   LOCAL RECOVERY

Global recovery schemes, such as DeCoR and the POWER6 processor's recovery unit, provide a coarse-grained recovery mechanism that can overpenalize parts of the processor that do not experience timing violations. In Section 1.1 we studied workload and unit-level variability. This motivates further solutions that can exploit workload and unit-level variability to reduce timing margins (more) aggressively.

Prior work proposed fully distributed local recovery mechanisms (LR) that are cognizant of interunit variability and thus reduce the overall recovery cost in the presence of emergencies. A fully distributed local-recovery mechanism entails overhead, particularly for the front end of the pipeline. Hence, there are two flavors of local recovery: (1) a fully distributed LR that completely eliminates the global recovery mechanism, and (2) a partial local recovery mechanism (PLR) that augments global recovery with local recovery for the execution units. The hardware required for implementing a local recovery mechanism is described below. An overview of global recovery is shown as the shaded logic in Figure 5.6. It is used to eliminate the global recovery unit.

Error Detection Unit (EDU)   Many server-class microprocessors now provide error-detection mechanisms distributed throughout the pipeline, often in the form of parity check, ECC, and residue codes [59]. More recent circuit-level schemes, as we saw earlier in this chapter (Section 5.1), can provide supplemental error detection [29, 94, 95]. When the EDU detects a violation, it triggers a recovery mechanism that simultaneously flushes the local pipeline and initiates the replay mechanism described below. For each execution pipeline in the processor, the detection unit is placed directly after the execution stage and before the delay and writeback stages. In the POWER6 pipeline, all units except the FPU have several delay-buffering stages following execution completion to ensure in-order writeback. These stages allow error detection to be performed off the critical path, in most cases with multiple stages before writeback. For the FPU pipeline, writebacks may need to be delayed depending on the timing of the EDU for FPU operations. Interpipeline communication (e.g., load-to-FXU dependencies) requires flushing both pipelines if the forwarding unit reports a violation, although this is not required for interpipeline dependencies through the register file. When an emergency is detected, the EDU sends a replay signal to the replay logic and a kill signal to the writeback stages of the execution pipeline. This prevents corrupted state from being propagated, which

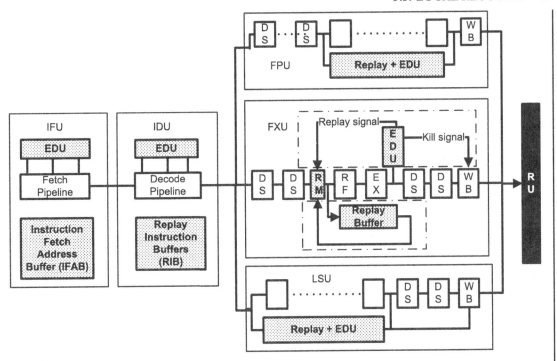

Figure 5.6: Architectural Support for Local Recovery. A baseline microarchitecture with architectural support for the local recovery mechanism. Additional hardware is shown by the shaded boxes. Details of the recovery mechanism for the execution units are presented for the fixed-point unit (FXU), and remaining units have the same logic represented as *Replay + EDU* logic. (DS: delay stages; RF: register file access; EX: execute; EDU: error detection unit; RM: replay mux.) Our distributed recovery mechanism replaces the recovery unit.

protects the architected state of the processor (the register files and memory units). The architected register file must be ECC-protected.

Replay Buffer for the Execution Units   Error detection must trigger a recovery mechanism. The unit in error needs to replay all operations performed on the internal, temporary state corresponding to the in-flight instructions in the unit's pipeline. Recovery is provided through replay buffers for each execution pipeline. Replay buffers are often used in processors to provide speculative execution of long latency or load instructions [62]. A similar replay mechanism can be used to provide recovery in the presence of timing-margin violations. The replay buffer stores the source and destination registers and the operation field for each instruction in the execution pipeline. The EDU sends a replay signal to the *replay mux*, which then feeds instructions into the execution pipeline from the replay buffer. A *stop* signal is sent to the scheduler, which halts scheduling of instructions to the recovering execution units; however, it is possible to continue scheduling nondependent instructions in the remaining execution pipelines. Normal scheduling resumes after all instructions are replayed. Figure 5.6 depicts the details of replay logic coupled with the EDU for the FXU; other units have similar logic, which is shown as a composite box (*Replay+EDU*).

Emergencies detected in the processor trigger a global throttling mechanism, which operates the processor in a low-frequency mode (at half the processor's frequency). This throttling mechanism both ensures that local restart will make forward progress and provides timing slack so that the logic in the error-recovery mechanism is free of timing errors. For the programs considered in this study, simulations show that a slow-restart period of 10 cycles is sufficient to guarantee forward progress.

Recovering the front end   Unlike the execution units, the processor's front-end units (fetch and decode) do not require replay buffers. The instruction stream fed into the back end is the relevant state that must be maintained in the front end. The fetch unit must save the instruction address buffer (PC-chain) to recover the fetched instruction sequence, which is saved in the *instruction fetch address buffers* (IFABs). The IFAB's size is proportional to the IFU's pipeline depth. Instructions fetched from the I-cache are written into instruction buffers in the IDU. To enable recovery of the IDU, a copy of the instruction buffers must be maintained (*replay instruction buffers* (RIB)). The instruction buffers can receive eight instructions per thread, every cycle. The RIB size is 8 times the number of stages in the IDU. It is important to note that all recovery state buffers (in both the front end and back end) require ECC protection, since, for functional correctness, these buffers must not be corrupted.

Implementation overhead   The fully distributed local recovery mechanism eliminates the need for a global recovery unit, distributing much of its functionality to the local recovery mechanisms. Most of the EDU logic is already present as part of the existing global recovery scheme. The overhead for local recovery is dominated by the ECC-protected replay state. However, for the back-end units, much of this state would already be present in an existing global RU, and hence, the state is simply being distributed across the local pipelines. In fact, for the LSU, most of the local replay logic and state may already exist in most processors to enable speculative execution for load instructions. The FXU pipeline requires a two-stage replay buffer (RF and EX stages), while the FPU pipeline requires an eight-stage replay buffer, due to the longer execution pipeline. In contrast to the back end, front-end units have higher replay state overhead. The IFU and IDU units deal with a wide stream of instruction and decode bits. We estimate that the RIB in the IDU requires eight instructions per pipeline stage in the IDU, leading to a 48-entry buffer per thread. While the total additional state required for a fully distributed recovery mechanism translates to less than 0.5 KB (less than 1% of the core latch count), the front end accounts for 75% of the overhead.

5.4   RAZOR

Circuit-level techniques provide fast response times with smaller recovery penalties as compared to the architectural-level schemes that we have discussed so far. Razor was the first circuit-level technique that proposed to lower processor voltage until the error-detection circuits detected voltage emergencies. More recently, such techniques have been tested in real chips, on Intel and ARM designs, and they have been demonstrated to work effectively. Intel-based chips demonstrated between 16% and 12% improvement [17]. ARM-based chips have shown that, for typical workload behavior, they can achieve as much as a 52% reduction in power while still maintaing correct execution [18].

Implementation   Razor [30] uses EDS-like circuits (Figure 5.1(a)) that are more tunable and flexible than TRC (Table 5.1), replicating flip-flops along the critical paths and sampling the output of the critical path twice to detect errors. In the simplest case, the EDS error signal can restore the incorrect value of the main latch using the value of the shadow latch. All the other flip-flops in the same pipeline stage must also replace their values at the same time. Sometimes, however, this is not possible, and more aggressive approaches are required to restore processor state.

The simplest, but slow, pipeline-level recovery solution is to use clock gating to stall the whole pipeline for a cycle. Whenever an emergency is detected in any stage of the pipeline, processor execution is halted by Razor (for a cycle). The global clock edge is gated, causing it to act as a buffering mechanism, and thus allowing every stage to get the result from its shadow latches. Operations can then safely resume with the correct value, which guarantees that the program makes forward progress, since the faulting instruction will simply continue execution in the next pipeline stage with the correct value.

In aggressively clocked processor designs, it may not be possible to implement global clock-gating without hurting processor performance significantly. Thus, the second technique is a counter-flow pipelined approach, whereby the faulting flip-flop distributes a bubble signal (forward) toward the end of the pipeline. At the same time, a flush signal is propagated (backward) toward the front of the pipeline. Thus, the notion of counterflow pipelining. The bubble moving down the pipeline ensures that the faulting instruction takes an additional clock cycle to complete its remaining stages, while the instructions following the faulting instruction are flushed out.

The flush control logic then restarts the pipeline at the instruction following the erroneous instruction. If multiple errors occur across the pipeline stages in the same cycle, then all stages initiate recovery, but only the error closest to the write-back stage will complete. All other earlier recoveries are automatically flushed out by later ones.

Aside from the conventional Razor techniques, two other techniques are based on replaying instructions for error recovery. The first is instruction replay at $\frac{1}{2} F_{CLK}$ [16, 17, 24]. The second is multiple-issue instruction replay at F_{CLK} [17]. Reducing F_{CLK} in half ensures that the replayed instruction executes correctly. When initiating the error recovery, the clock generator reduces F_{CLK} in half, which can be done via a clock-divider circuit to avoid relocking of the phase-locked loop (PLL). While F_{CLK} reduces in half, the duty-cycle control circuit maintains a constant high-phase clock delay to provide min-delay protection for the embedded EDS design. After the replayed instruction finishes, the ECU signals the clock generator to resume at the target F_{CLK}.

The motivation for the multiple-issue instruction replay design is to correct the errant instruction without changing F_{CLK}. This technique issues the errant instruction multiple (N) times. The first $N - 1$ issues are replica instructions that do not affect the architecture state. The Nth issue is a valid instruction, which is allowed to commit data to the architectural state. The replica instructions flow through the pipeline to set up the register nodes for the valid instruction. Any error that occurs in the execution of the replica instructions is ignored, and if the number of replica instructions is sufficient, the register inputs for each pipeline stage statically settle to the correct value. This allows the valid instruction to execute correctly. If an insufficient number of replica instructions is issued such that an emergency occurs during the execution of the valid instruction, then the errant instruction is replayed a 2nd time with N equal to the number of pipeline stages to guarantee correct operation. Since this error-recovery design relies on setting up path nodes, this technique is directly applicable to static-CMOS logic circuits. This technique is not applicable to dynamic logic circuits.

Benefits   To illustrate the benefits between an EDS- and TRC-based Razor implementation, here we discuss the results of a 45nm microprocessor processor core [17]. In Figure 5.7(a), the core without resilient circuits (i.e., conventional design) operates at a maximum clock frequency (F_{MAX}) of 1.45 GHz at 1.0V. When a 10% V_{CC} droop occurs, the F_{MAX} reduces to 1.26 GHz. As illustrated by the shaded region in Figure 5.7(a), the difference between these two F_{MAX} values represents the F_{CLK} guardband for a 10% V_{CC} droop in the conventional design. Enabling EDS or TRC allows the infrequent errors from the V_{CC} droop to be detected and corrected, resulting in a higher F_{CLK} and throughput (TP). The optimal F_{CLK} for the resilient designs (1.46 GHz for EDS, 1.42 GHz for TRC) occurs at the point of maximum TP. Pushing F_{CLK} beyond this point reduces TP because the increasing number of recovery cycles outweighs the benefit of a larger F_{CLK}. In comparison to the conventional design, EDS and TRC circuits improve TP by 16% and 12%, respectively, at 1.0V.

EDS vs. TRC Trade-Off   Although an EDS design provides a larger benefit at 1.0V, the error-detection window, and corresponding potential TP gain, is limited by core min-delay constraints (Section 5.1). In contrast, the core min-delay constraints do not limit the error-detection window for the TRC design, which allows the TRC design to capture a wider range of dynamic delay variation. In Figure 5.7(b) at low V_{CC}, the impact of variations increases and TRC provides more throughput gain than EDS (51% vs. 28% at 0.6V).

5.5   SUMMARY

Circuit-level embedded error correction incurs major penalties for detectors and correcting control loops. Generally, speaking, a 10% area penalty can provide a 40% improvement in performance, or a 20% reduction in energy [53]. With 30% transistor-level performance improvement each technology generation, the question that circuit designers and architects must address is whether the 40% circuit-level performance improvement is worthy of all the added area, design, and validation complexity introduced by resiliency mechanisms.

Circuit-level error correction provides very low latency, but it is hard to embed in optimized designs. Specifically, circuit-level techniques may be too costly to implement in a modern high-performance, out-of-order core with several large array structures and tight timing paths. These

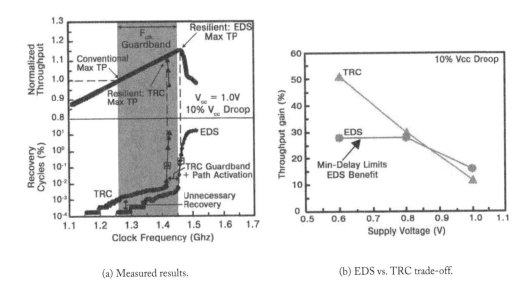

(a) Measured results.

(b) EDS vs. TRC trade-off.

Figure 5.7: Performance Benefits. (a) Measured throughput normalized to the conventional maximum throughput, and recovery cycles as a percentage of total cycles vs. F_{CLK} [17]. (b) Measured throughput gain vs. V_{CC} [17], showing that TRC can outperform EDS-based error detection for Razor-type design.

circuit-level techniques must be applied carefully to all the critical executions paths for timing violations. Annavaram et al. showed that the distribution of timing margins for different paths across functional blocks in the Intel Core Duo microprocessor have hundreds of paths within 10% timing margins [5]. Designs such as Razor impact overall processor performance and area overhead, as well as design and verification effort. For instance, Razor is estimated to add between 1% and 3% area overhead, and has an estimated penalty between 1% and 3% during normal conditions.

In general, timing and completion uncertainties caused by embedded error-correction techniques pose added challenges in design verification, validation, and test. All circuit types and arrays must be covered with resilient solutions to make a difference in overall power-performance trade-off. Designers must evaluate the techniques across a spectrum of practicality issues in the real world. Microarchitectural level correction, such as DeCoR, is an alternative, but it may not work for all types of machines, e.g., in-order vs. out-of- order execution pipelines. DeCoR implicitly relies on value buffering of the noise-speculative state in the reorder buffer (ROB) and store queue (STQ), which are specific to out-of-order processor design. Moreover, it adds to the processor's design complexity.

In other words, there is no magic bullet that specifically addresses the problem without penalties. There is a need for an end-to-end platform solution. Industry is moving toward a collaborative architecture design, involving some level of tolerance combined with proactive resiliency techniques at the hardware (and the software) layers. In this chapter, we discussed only reactive solutions that detect and respond to an emergency. Going forward, we require proactive solutions that anticipate an emergency and take preventive measures to balance the penalty trade-offs intelligently, based on the emergencies characteristics.

CHAPTER 6

Predicting and Avoiding Voltage Emergencies

Possibly the first microarchitectural-level approach to reducing voltage variation was by Pant et al. [69], clearly identified that sudden swings introduced by clock gating in the microprocessor were the primary reason for large voltage variation. Their recommended approach was to gradually activate and deactivate functional units as required to mitigate the large and sudden current fluctuations. Over recent years, more effort has been focused on investigating other architecture-level solutions to *avoid* voltage emergencies [35, 50, 72, 74].

In general, many of the techniques, unlike tolerance, strive to avoid impending voltage emergencies arising from voltage variation to prevent failures. Avoidance mechanisms require proactive detection of impending voltage emergencies, not reactive detection as we discussed in the Chapter 5. In this chapter, we discuss these sensor- and event-based detection mechanism. On the basis of detecting an impending emergency, we discuss a plethora of avoidance throttling mechanisms, such as pipeline muffling and phantom firing.

6.1   SENSOR-BASED THROTTLING

Joseph et al. [50] propose a voltage-sensor-based approach, where a throttling mechanism is invoked when the sensed supply voltage crosses a specified level, called the soft threshold. As shown in Figure 6.1, the sensor turns on a throttling mechanism (i.e., the actuator). The hardware-based actuation mechanism responds with a precautionary measure whenever the source voltage moves outside of the predefined control range. As the source voltage moves outside of the control threshold range, the processor reacts by performing one of two actions. If the alarm is resulting from a sudden reduction in current draw, the processor responds by producing phantom firings of one or more functional units. Furthermore, if the current suddenly increases, the processor disables one or more functional units. Although these techniques effectively correct impending voltage emergencies, the latter case does so at the expense of program performance, while the former case wastes energy.

Sensor-based techniques are becoming mainstream. A sensor-based throttling scheme called Critical Path Monitoring (CPM) has been implemented in the IBM POWER7 server. CPMs measure the available timing margin dynamically and adjust the operating voltage to maintain a fixed-timing guardband determined during worst-case characterization. The resulting mechanism reduces power consumption for typical workloads, while still allowing worst-case workloads to operate at the maximum frequency used in the characterization process. Lefurgy et al. [58] show

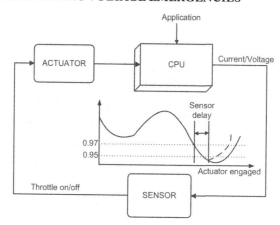

Figure 6.1: Sensor-Based Emergency Avoidance. The feedback loop that is associated with any sensor-based emergency-avoidance mechanism.

that during better-than-worst case conditions the average voltage automatically reduced by137 to 152 mV below nominal, which resulted in a 24% reduction in power consumption without any impact on performance.

6.1.1 DESIGN

Actuator   In general, there are several different ways to reduce current variations via throttling mechanisms. These throttling mechanisms include frequency throttling [35], phantom firing [50], pipeline muffling [74], pipeline damping and a priori current ramping [73], and changing the number of the available memory ports. Throttling simply reduces the processor's clock rate by a certain amount, say by half, in order to slow down the amount of processor current draw. Pipeline muffling reduces current variability either in space (i.e., variability in usage of circuit blocks) or in time (i.e., variability within a circuit block across clock cycles). It does this by controlling instruction issue and limiting changes in the number of resources utilized. Pipeline damping gradually raises resources' current draw over time to reduce the rate of sudden current change in individual resources. A priori current ramping allows time for the resource current to gradually ramp up a few cycles ahead of utilization and to ramp down immediately after utilization. Resonance tuning is yet another throttling mechanism that assumes, and specifically targets, voltage emergencies caused by repeating high-low/low-high current transitions, occurring within the resonance band [72]. Phantom firing is the process of activating idle functional units whenever there is a sudden reduction in current draw. Most of these approaches attempt to control current draw, but, in practice, they end up impacting application performance because they typically slow down the machine.

An alternative to the above performance throttling mechanisms is expending energy in an execution-independent manner. Alon and Horowitz [2] propose a push-pull regulator topology that

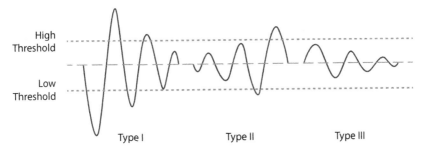

Figure 6.2: Warning Types. Classification of various voltage variation levels.

uses an additional higher-than-nominal supply voltage, comparator-based feedback, and a switched-source follower output stage to reduce voltage variation. This regulator's output stage can be used as an actuator. This technique has an effect similar to throttling, but without the extra performance penalty associated with throttling.

All of these mechanisms rely on voltage or current sensors to detect threshold crossings indicating timing-margin violations having occurred or about to occur. As shown in Figure 6.1, the sensor turns the throttling mechanism (i.e., the actuator) on. Assuming that the control logic and actuation mechanism can react quickly, the main bottleneck in throttling for emergency avoidance is the speed of the individual sensors and aggregation across multiple sensors in different parts of the processor. This delay sets the speed of the overall feedback loop. There are many ways to build sensors with tradeoffs between delay and precision. Hence, it is important to understand the impact of the inherent delay and inaccuracy associated with the sensors on the different emergency-avoidance schemes.

Sensor A throttling mechanism must react before the voltage deviation proceeds beyond the soft threshold to the hard threshold. The sensors must have the capability of tracking voltage to trigger various warning levels. It is important to note here that not all warnings constitute an emergency. Recall that voltage variation can arise due to the interaction between parasitic inductance and capacitance in the power-delivery network during events that cause sudden fluctuations in the current (i.e., $\frac{di}{dt}$). In light of the power-supply network's resonance, the warnings can be broken up into various categories.

Figure 6.2 illustrates the different types of warnings that sensors typically need to identify. The first class (Type I) corresponds to a case when a large $\frac{di}{dt}$ event occurs that causes large voltage fluctuations, which immediately exceed prescribed high and low thresholds required to guarantee proper circuit operation. This warning signifies an emergency that requires immediate action. The second class (Type II) corresponds to a warning scenario where the initial $\frac{di}{dt}$ event does not constitute an immediate emergency. However, continued occurrence of the event at or near the resonant frequency of the power-supply network will eventually lead to catastrophic failure and, therefore, will require action at some point. The third class (Type III) presents a scenario where the power-supply voltage is perturbed, but is eventually damped by the PDN and does not require action.

Given these different scenarios, the best throttling solution to address the problem can differ. A Type I alarm requires a hardware solution that responds hard and fast. For a Type II alarm, a responsive but slowly reacting solution may be possible, depending on how long it takes for the warning to become a catastrophic emergency. Clearly, a Type III warning does not constitute an emergency and requires no action.

The warning classifications highlight desirable sensor attributes that must be in place for sensor-based throttling schemes to operate. Instead of simply detecting when the supply voltage crosses preset thresholds, it may be able to detect the rate of change of the voltage fluctuations. This additional information can enable the sensor subsystem to track the trajectory of voltage signals in order for the microarchitecture to better distinguish the class of a warning, predict when upcoming emergencies will occur, and take action. This methodology for sensing emergencies is especially useful since hardware-level actions taken to handle emergencies can mask the downstream consequences of coupled events.

One can envision multiple ways to construct sensors. Here, we briefly describe the CPM circuit that is used in the POWER7 processor [26]. The CPM is a critical path synthesis circuit. The CPM relies on a variety of delay paths that are constructed using a mix of field-effect transistor (FET) and wire delay to approximate the critical paths that dominate within the microprocessor over the processor's range of operating frequencies.

Figure 6.3 is a high-level block diagram of a CPM circuit. In order to take a measurement, a pulse is sent down the CPM circuit's delay paths at the start of clock cycle n and the penetration of the pulse through the delay paths is captured in a 12-bit edge detector on the following clock cycle, $n + 1$. The amount of penetration of the pulse into the 12 bits gives an approximation of the circuit's timing margin at the current operating point of the processor. Note that the pulse's delay through the synthesis paths is in fact a function of different processes, such as voltage, temperature, workload, aging, etc. It is not just a voltage variation sensor. Thus, the CPM provides a measurement of the chip's operating environment on any given clock cycle. It effectively allows us to measure how the amount of variation in the environment is affecting the current timing margin of that region of the chip. Several CPMs are required to observe the global state of the processor.

Also, it is worth noting that designers must understand the sensitivity of sensors, such as CPMs, to noise caused by temperature, voltage, and clock frequency. The circuits must be calibrated accordingly across a range of operating environment characteristics.

Feedback Loop   A typical sensor-based proposal uses a tight feedback loop like that shown in Figure 6.1. The loop includes a sensor on the critical path that tries to detect impending emergencies to actuate the throttling mechanism based on the soft threshold level. Assuming that the control logic and actuation mechanism can react quickly, the main bottleneck in throttling for emergency avoidance is the speed of the individual sensors and aggregation across multiple sensors in different parts of the processor. This delay sets the speed of the overall feedback loop, which in turn implies

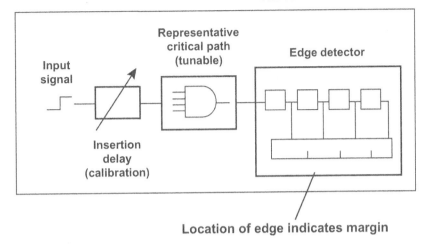

Figure 6.3: Block Diagram of a Critical Path Monitor Circuit. Signal penetration depth into the edge detector indicates how much timing margin exists for the chip's circuitry to operate.

that the choice of the soft threshold level is largely governed by the voltage-sensor response time and accuracy. There are many ways to build sensors with tradeoffs between delay and precision. Hence, it is important to understand the impact of the inherent delay and inaccuracy associated with the sensors on the different emergency-avoidance schemes.

6.1.2 CHALLENGES

We previously emphasized the importance of a power-delivery network's Q factor while designing any technique to handle voltage emergencies (Section 3.2). Specifically, the rate change of voltage will depend on Q. For example, a snapshot of *crafty's* voltage trace is depicted in Figure 6.4(a). In this example, we assume two thresholds, one at ±3% (soft) and the other at ±5% (hard). The short time interval over which the voltage crosses both soft and hard thresholds determines the maximum delay any soft-threshold-based avoidance mechanism can tolerate. Figure 6.4(b) plots the percentage of voltage emergencies occurring for different delays between threshold crossings over a range of Qs across the entire voltage trace of *crafty*. This plot shows that, even with moderately low Qs, voltage fluctuations can be very fast.

Sensor Delay   Figure 6.4(a) shows that it takes only three cycles from the crossing of the ±3% soft threshold to the crossing of the ±5% hard threshold for the program *crafty*. In general, the maximum allowable sensor delay is largely determined by the minimum number of cycles for voltage to transition between the two thresholds. This suggests that for the emergency-avoidance throttling mechanism to work correctly for *crafty* under the given package model, the sensors must detect the soft-threshold crossing within two cycles, leaving only one cycle for the actuator mechanism.

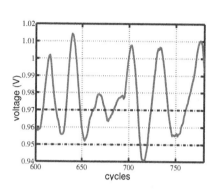

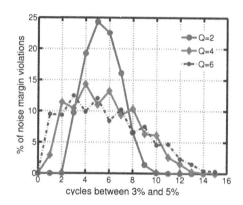

(a) A snapshot of the voltage profile, assuming a package with Q=2.

(b) Number of cycles over which the voltage crosses 3% and 5% thresholds.

Figure 6.4: Need for Fast Sensors. A voltage-emergency snapshot of a SPEC CPU2000 program—*crafty*. Figure 6.4(a) shows the cycles between the crossing of a threshold of 3% and 5%. Figure 6.4(b) shows the distribution of these cycles for packages with different Q factors.

Because voltage emergencies are rare events, one might argue that the fraction of those occurring with such a steep slope would be extremely rare. Unfortunately, there needs to be a single timing-margin violation to disrupt the reliability of the processor circuits and cause incorrect program execution. Consequently, all such situations must be avoided, and as such, the designer should carefully calibrate the soft threshold accordingly to allow enough lead time to avoid the emergency.

Threshold Setting   Figure 6.5 illustrates the use of a soft threshold to throttle execution and prevent an emergency. The figure shows a voltage profile snapshot with and without sensor-based throttling. The dotted horizontal lines marked *Soft Threshold* indicate the thresholds at which a sensor-based scheme takes action to prevent an emergency. Setting the soft threshold aggressively (i.e., close to the timing margin, as illustrated in Figure 6.5(a)) requires fast reaction by the sensor and actuation system. Otherwise, failure to respond quickly results in a voltage emergency, leading to correctness violations. The original execution waveform in Figure 6.5(a) represents the voltage activity of a system that is operating without a mechanism to prevent emergencies. Left uncorrected, the voltage exceeds the lower operating margin. Throttled execution is the voltage activity under a sensor-based throttling mechanism to prevent emergencies. With a sensor-based throttling mechanism in place, the sensor actuates the throttling mechanism upon crossing the soft threshold. The throttled-execution voltage waveform starts recovering, but not in time to avoid exceeding the lower operating margin.

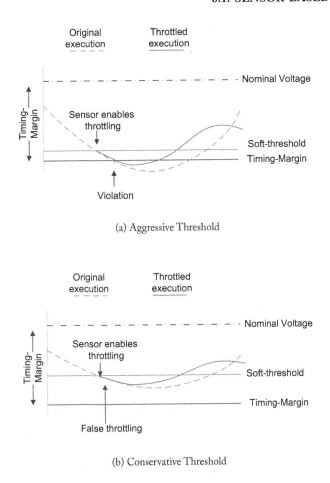

(a) Aggressive Threshold

(b) Conservative Threshold

Figure 6.5: Choice of Thresholds. The impact of choice of soft thresholds on the performance and correctness guarantees. (a) Aggressive thresholds allow too little time to prevent emergencies, which leads to violations. (b) Conservative thresholds increase the rate of unnecessary throttling, leading to performance loss.

To guarantee correct operation, emergency-avoidance throttling mechanisms must apply the throttling mechanisms before an emergency actually develops. To provide the sensor and actuator with more time to operate when delays are long, one can consider increasing the distance between the timing margins and soft thresholds. However, conservatively setting the soft threshold increases the number of false alarms, where voltage variations are unnecessarily flagged as requiring throttling. Figure 6.5(b) illustrates how false alarms can arise with conservative soft thresholds. The original execution voltage waveform does not fall below the lower margin. But a sensor-based scheme using a

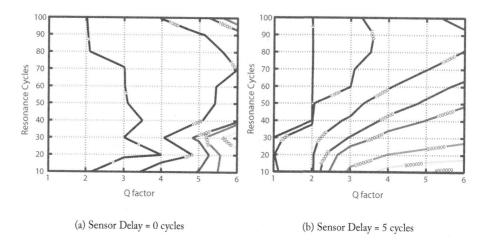

(a) Sensor Delay = 0 cycles   (b) Sensor Delay = 5 cycles

Figure 6.6: Sensitivity to Package and Sensor Delays. The number of timing-margin violations for SPEC program *bzip* with different package solutions and different sensor delays.

conservative soft threshold incorrectly assumes an emergency is about to occur and throttles execution to no benefit. Not every soft threshold crossing goes on to exceed the operating margin, and for this reason it is possible to throttle unnecessarily and degrade performance.

Robustness   A processor's susceptibility to voltage emergencies is tightly coupled to the underlying power-delivery network, as we discussed in Section 3.2. Hence, any proposed solution's correctness depends on assumptions made about the package and/or processor models. We show that the throttling schemes' correctness varies with respect to current swings (dependent on processor architecture), resonant frequencies (i.e., packaging assumptions), and sensor delays. To illustrate this point, it is sufficient to sweep two of the three parameters governing power-delivery network characteristics—Q factor and resonance cycles—as changing either one affects the peak impedance of the system.

The techniques discussed in Section 6.1 avoid emergencies by reducing current fluctuations by throttling the system in response to detection of voltage [50] or current repetitions [72]. Figure 6.6 presents contour plots of the number of timing-margin violations for *bzip* across different package characteristics, while employing an aggressive, 0.5× frequency-throttling mechanism that responds to voltage swings. With a voltage-sensor soft threshold of ±3%, consider two sensor-delay scenarios—a sensor delay of 0 (Figure 6.6(a)) and a sensor delay of 5 (Figure 6.6(b)). Even with an optimistic sensor delay of 0, this throttling scheme fails to prevent timing-margin violations for packages with Q greater than 2, leading to correctness violations. For more realistic sensor delays, the number of violations increases by two orders of magnitude, and even packages with relatively low Q are sometimes unable to avoid timing-margin violations. The package characteristics as-

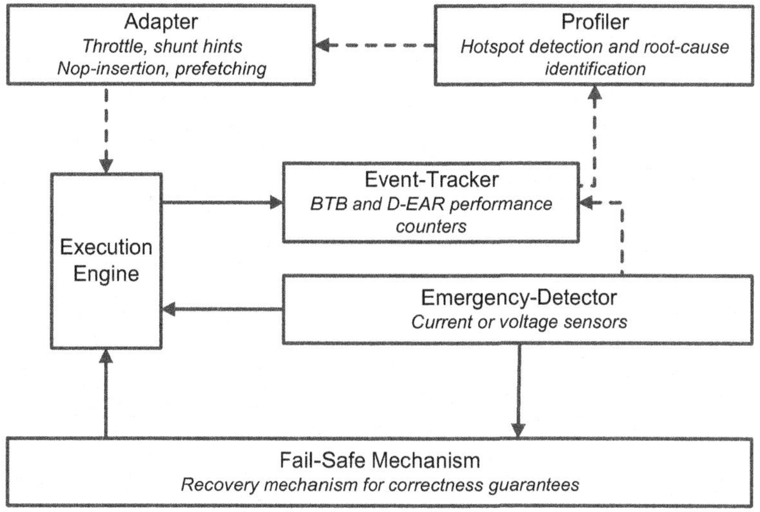

Figure 6.9 area:

Adapter
Throttle, shunt hints
Nop-insertion, prefetching

Profiler
Hotspot detection and root-cause identification

Execution Engine

Event-Tracker
BTB and D-EAR performance counters

Emergency-Detector
Current or voltage sensors

Fail-Safe Mechanism
Recovery mechanism for correctness guarantees

Figure 6.9: Event-Guided Predictor Architecture. A feedback-driven voltage-emergency prediction architecture based on program and architectural events for handling emergencies.

must roll the machine back to a known correct state and resume execution. Subsequent occurrences of the same emergency signature cause the predictor to throttle execution and prevent the impending emergency. By doing so, the predictor can effectively enable aggressive timing margins to maximize performance, even in the presence of emergencies, but without compromising performance unnecessarily. The cost of signature-based throttling is fewer than 10 cycles, much less than the cost of recovery.

Figure 6.10 illustrates example snapshots of emergencies, and the corresponding processor activity history that is captured. At the point of the emergency during execution region B, the history of activity contains (from oldest to most recent) two control flow instruction addresses (illustrated as BR) and an event encoding for the pipeline flush (illustrated as 2), followed by another branch. The function of a voltage-emergency signature is to precisely indicate whether a pattern of control flow and microarchitectural event activity will give rise to an emergency. To this end, the accuracy of a signature, where accuracy is defined as the fraction of predicted emergencies that become actual emergencies during execution, is a measure of its contents and size.

Contents Information tracked in the signature must correspond to parts of the execution engine that experience large current draws, as well as dramatic spikes in current activity. The signature can collect the control flow trace at different points in a superscalar processor: in-order fetch and decode, out-of-order issue, and in-order commit. Each of these points contributes different amounts of information pertaining to an emergency. For instance, tracking execution in program order fails to capture any information regarding the impact of speculation on voltage emergencies. Tracking

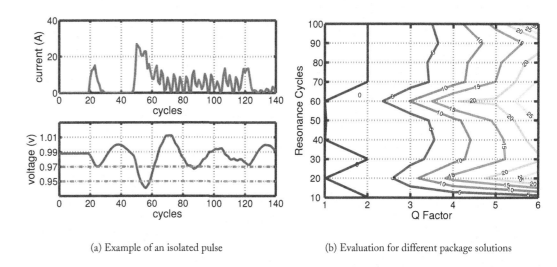

(a) Example of an isolated pulse

(b) Evaluation for different package solutions

Figure 6.7: Evaluation of Resonance Tuning Technique. An evaluation of the resonance tuning scheme [72] for *equake*. (a) depicts the current and voltage profile for few hundred cycles for *equake*. (b) depicts the percentage of single transition events causing timing-margin violations for different package solutions.

sumed in [50] lie in the small region (upper left corner) where the throttling mechanism is effective. These results show that a throttling-based emergency-avoidance scheme alone is not robust enough to be a general solution to guarantee correctness, and therefore, care must be taken in its design, implementation, and application.

Caution must be taken in using specialized avoidance mechanisms at very aggressive margins. For instance, resonance tuning specifically avoids voltage emergencies that are caused by repeating high-low/low-high current transitions in the resonance band [72]. At aggressive margins, in addition to resonant pulses, several single-transition events can occur, and any one of these events can cause timing-margin violations. Unfortunately, resonance tuning cannot cope with single-transition events, which can lead to erroneous execution. For example, Figure 6.7(a) illustrates an example of a sudden isolated current pulse found in SPEC workload *equake* that causes voltage to swing below the hard threshold. Figure 6.7(b) shows a contour plot delineating the fraction of single-transition events (or isolated pulse emergencies), seen in *equake*, that caused timing-margin violations across different package characteristics. These results show that resonance tuning by itself would not be able to detect such pulse emergencies for packages with Q greater than 2.

6.2   EVENT-BASED THROTTLING

Sensor-based predictors operate independently of program or microarchitectural state. Higher-level information has the intrinsic property that it relates program/machine activity to power supply behavior. The intuition behind this relationship is that processor-current draw depends on the set of functional blocks that are active and consuming power during each cycle. The activity of these functional blocks depends on the set of instructions in flight through the core's pipeline, thus relating current draw and consequently voltage flux to higher-level program instruction sequences.

6.2.1   SINGLE-EVENT PREDICTORS

An approach to eliminate the challenges of feedback-loop delay associated with sensor-based techniques is to instead monitor specific microarchitectural events as indicators of processor activity that can lead to voltage emergencies. In effect, this *predictor* replaces current and voltage sensing in Figure 6.1 with microarchitectural event detection. Such an event-driven mechanism triggers corrective action when it detects certain emergency-prone events (L2 cache misses and branch flushes, as they are the events associated with most of the emergencies). Certainly, a naive implementation might take preventive measures at every such event (e.g., to activate a throttling mechanism at every L2 miss). That would be overly conservative, however. Because most such events do not give rise to emergencies, such a system lends itself to a high false-alarm rate of 71% [36].

Instead, by tracking specific instructions associated with events (L2 misses or pipeline flushes) that have caused emergencies, and maintaining contextual information for each event and emergency, it is possible to reduce the false rate significantly. Reacting only to events associated with emergencies results in much less overhead than the naive implementation. Figure 6.8 shows a cumulative distribution graph plotting the number of unique program addresses that trigger emergencies and their contribution to the total number of emergencies during execution. Each program except for *parser*, *gcc*, *twolf*, and *crafty* has fewer than 15 unique program addresses that cause over 90% of runtime emergencies.

The challenge of event-based prediction is finding the leading indicator (i.e., unique program addresses). It requires a method that learns to avoid recurring voltage emergencies by triggering preventive action on the microarchitectural events that cause them. Figure 6.9 shows an overview of the architecture for the event-guided predictor to detect and suppress voltage emergencies. The parts of the diagram that are connected by solid arrows detect and respond to actual voltage emergencies. The parts connected by dashed arrows are responsible for learning to recognize impending violations and using this training to suppress future occurrences of violations. Current and voltage are monitored by a sensor, and an emergency handler determines when the supply voltage exceeds operating margins. Note that, since the event-guided predictor is a *reactive* mechanism that learns by allowing an emergency to occur, it is not subject to the same issues as sensor delay, discussed previously in Section 6.1. Upon detecting a voltage emergency, the handler invokes a fail-safe mechanism to recover from any deleterious effects of an emergency. This fail-safe mechanism can be any one of the recovery mechanisms we described earlier in Chapter 5. The handler also signals an event

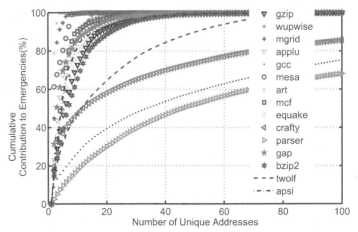

Figure 6.8: Single Event-based Voltage Emergency Prediction. The number of unique instructions causing emergencies and their corresponding contribution to the total number of emergencies [36].

tracker to learn from this emergency, in order to recognize future emergencies, which passes the relevant information to the triggering layer. Later, when the triggering layer detects an emergency-prone situation, it invokes the adaptation mechanism to take appropriate preventive action. The adaptations can be any one of the techniques borrowed from the sensor-based throttling schemes.

While there are just a few instructions, prior research demonstrated that event predictors are a poor heuristic for whether voltage emergencies are likely in the next few clock cycles [78]. Prediction accuracy can be poor. Single-event prediction accuracy is only about 10%. As we discussed in the tolerance section, the history of activity leading to voltage emergencies is important. A single-event predictor does not capture sufficient history for accurate emergency anticipation. Gupta et al. [36] also confirm these results, showing that throttling is a poor adaptation mechanism for the event-guided scheme described here.

6.2.2   SIGNATURE-BASED PREDICTION

To overcome the limitations of event-based predictors, another approach that has been proposed is a voltage-emergency predictor that identifies when emergencies are imminent and prevents their occurrence by predicting them using signatures [78]. An emergency predictor predicts voltage emergencies using emergency signatures and throttles machine execution to prevent them. An emergency signature is an interleaved sequence of control-flow events and microarchitectural events leading up to an emergency.

A voltage-emergency signature is captured when an emergency first occurs (tolerated) by taking a snapshot of relevant event history and storing it in the predictor. Tolerance is an integral part of this predictor because, after detecting an emergency, a fail-safe checkpoint-recovery mechanism

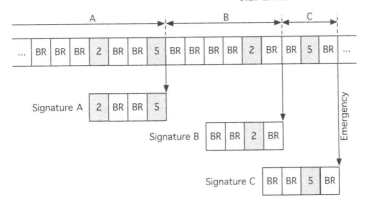

Figure 6.10: Voltage-Emergency Signatures. Taking snapshots of recent processor microarchitectural and program activity for an example set of three voltage emergencies.

information at the in-order fetch and decode sequence captures the speculative path, but it does not capture the out-of-order superscalar issuing of instructions.

The accuracies of different signature types are illustrated in Figure 6.11(a) (assuming a signature size of 32 entries, which will be discussed next). Tracking committed control-flow sequences gives an accuracy of only 40%. If information is tracked at the decode stage, an accuracy of 72% is possible because the decode stage captures the speculative control-flow path. Accuracy improves further by 12%, from 72% to 84%, if control flow is tracked at the issue stage, since we can now capture interactions more precisely at the level of hardware instruction scheduling and code executed along a speculative path.

Interleaving microarchitectural events with program control improves accuracy even further, as processor events provide additional information about swings in the supply voltage. For instance, pipeline flushes cause a sharp change in current draw as the machine comes to a near halt before recovering on the correct execution path. The last two bars of Figure 6.11(a) show the accuracy improvements from adding microarchitectural event activity. The second to last bar represents the effect of capturing events that have the potential to induce large voltage swings—pipeline flushes and secondary (L2) cache misses. An improvement of five percentage points is achieved by taking flushes and L2 misses into account (i.e., total accuracy of 89%). Capturing the more frequently occurring events like DTLB and DL1 misses contributes additional improvements of ~4%. Microarchitecture perturbations resulting from instruction cache activity (i.e., IL1 and ITLB) have been shown to be negligible, and thus, do not lead to noticeable improvements in accuracy.

Size   Assuming signature information is captured at the issue stage of the pipeline, accuracy also depends not only on recording the right interleaving of events, but also on balancing the amount of information kept in the voltage-emergency signature. Typically, accuracy improves as the length of signatures increases. However, it can be detrimental to increase the number of register entries beyond

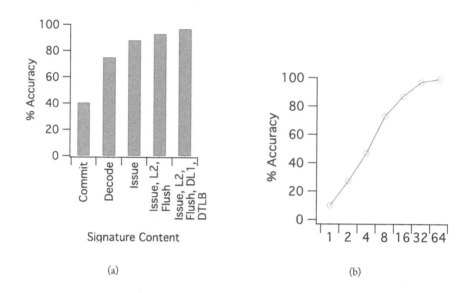

(a)

(b)

Figure 6.11: Signature-based Prediction Accuracy. Prediction accuracy improves as (a) signature contents represent machine activity more closely and as (b) the number of entries per signature increases.

a certain count. Large numbers of entries in a signature can cause unnecessary differentiation between similar signatures—signatures whose most recent entries are identical and whose older entries are different, but not significantly so. The predictor would have to track more unique signatures per emergency because of this differentiation.

Figure 6.11(b) shows prediction accuracy improves as signature size increases. Accuracy is only 13% on average for a signature containing only 1 entry, which supports the discussion presented in Section 6.2 whereby Gupta et al. discovered that voltage emergencies do not depend solely upon the last executed branch or a single microarchitectural event [36]. It is the history of activity that determines the likelihood of a recurring emergency. Prediction accuracy begins to saturate once signature size reaches 16, and peaks at 99% for a signature size of 64 entries. However, accuracy is to be taken with caution here, since we must also think about the hardware requirements to support that level of accuracy.

Lead Time   For a voltage-emergency predictor to be effective, the predictor must anticipate an emergency accurately and do so with sufficient lead-time for throttling to take effect. Signatures predict emergencies with an average accuracy of 93% across the entire spectrum of SPEC CPU2006 programs, as illustrated via the 0-cycle lead-time bar in Figure 6.12. A lead time of 0 cycles optimistically assumes there is no delay to actuate throttling to prevent an emergency, thus representing an ideal scenario. However, real systems require non-zero lead times to account for circuit delays,

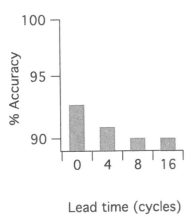

Figure 6.12: **Lead-time Prediction**. Prediction accuracy of a signature-based predictor is high even when predicting cycles ahead of time [78].

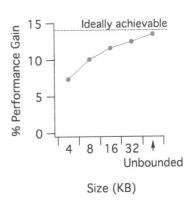

Figure 6.13: **Performance Benefits**. A signature-based predictor enables substantial gains using an aggressive 4% voltage margin [77].

as we discussed previously in sensor-based schemes. Naturally, as lead time increases, Figure 6.12 shows that accuracy degrades from 93%.

Throttling cannot prevent all emergencies, even if prediction accuracy is high. In such cases, the fail-safe recovery mechanism recovers processor state, albeit at much higher penalty. But the number of such emergencies is typically only 1% of all emergencies that occur without throttling. Therefore, total recovery penalty observed is typically low.

An aggressive reduction in operating voltage margins translates to higher performance or better energy efficiency. However, benefits are offset to some degree because of throttling penalties to prevent emergencies and checkpoint-recovery rollbacks to train the predictor. In simulations of a representative superscalar microprocessor in which fluctuations beyond 4% of nominal voltage are treated as voltage emergencies, a signature-based predictor shows great promise. Based on a 1.5× relationship between voltage and frequency at the PTM 32nm node [99], we observe an ideal performance gain of 14.2% using an oracle throttling scheme (see Figure 6.13). By comparison, the voltage-emergency predictor comes to within 0.7% points of the ideal scheme, assuming infinite or unbounded resources to implement the predictor. But even under strict physical resource constraints, an intelligent bloom filter-based predictor ranging in size between 4 KB and 32 KB delivers substantial gains.

Robustness   An added benefit of signature-based emergency prediction is that the predictor does not require fine tuning based on specifics of the microarchitecture or the power-delivery network, as is the case with sensor-based predictors. The current and voltage activity within a microprocessor are products of machine utilization that are specific to running workload-dynamic demands. Capturing that activity in the form of emergency signatures allows the predictor to dynamically adapt to the emergency-prone behavior patterns resulting from the processor's interactions with the power-delivery network without having to be preconfigured to reflect the characteristics of either.

6.3 SUMMARY

Avoidance is an important first step in proactive architectural solutions to the voltage variation problem. In this chapter, we examined microarchitectural control mechanisms for predicting the problem using a variety of mechanisms, ranging from sensor-based thresholds to using microarchitectural event(s). These architectural avoidance mechanisms assume operation with and without backing support to restart and replay instruction execution, providing a tradeoff between strictly proactive strategies that have limitations on the extent of guardband margin reductions, and partially proactive schemes, which, when combined with reactive strategies, enable more aggressive margin reductions.

The choice of implementation, however, is not a straightforward and obvious one. The strength and duration of voltage transients vary depending on the characteristics of and interactions between the power-delivery network, the processor and runtime workload activity. In general, it seems a proactive mechanism combined with a coarse-grained recovery scheme offers a well-balanced system that is capable of dynamically learning and avoiding the interactions that lead to emergencies in the field.

Despite advances at the architecture layer, there is a need for an end-to-end system-level solution for resiliency. Industry is moving toward hardware-software co-design. While software-based correction methods are typically slow to react, the performance/energy gains they offer are substantially high (assuming the software overheads can be amortized over runtime). Software-driven improvements are likely to be in the order of multiples. Moreover, they are flexible. Their sensitivity and intrusiveness can be dynamically calibrated based on a system's runtime requirements. In contrast, hardware design needs to be well thought-out beforehand, requiring costly design and testing up-front. Application adaptation and/or error tolerance is also possible at the software level. For instance, some level of error allowance can be linked with application- and/or user-driven instruction criticality tags at compilation. So, we conclude that hardware should include sensors, environment monitors (or circuit-level error detectors) that are coupled with sub-optional architecture-level reactive and proactive measures that ultimately enable error correction at a higher level of abstraction.

CHAPTER 7

Eliminiating Recurring Voltage Emergencies

Hardware-based solutions typically work well for intermittent voltage emergencies, but a loop incurring repeated voltage deviations may be handled more gracefully by a compiler. A hardware-based solution may repeatedly throttle, or roll back, on an emergency recurring at the same program location because it lacks global knowledge involving program structure and activity. In contrast, a compiler typically has several options when choosing the order of instructions, and many of these options result in equally performing software. Therefore, in the case of the voltage-emergency loop, the compiler may be able to rearrange the instructions to avoid the voltage emergency without impacting performance.

Moreover, in the context of multicore systems, coordinating hardware thread activity across cores is substantially more challenging because of issues such as wire lengths and propagation delays. It is difficult to simultaneously observe chip-wide activity at the hardware level to make coordinated decisions at the individual core level. And as such, intelligently scheduling threads at the software level to dampen voltage swings is likely to become important because of its feasibility and practicality. As the number of processor cores sharing a power-supply source increases, the absolutely peak-to-peak voltage swings may also increase due to interfering microarchitectural activity across hardware contexts.

Hazelwood and Brooks [40] showed that voltage emergencies are correlated with an application's dynamic code stream and not just the underlying architecture and power-delivery subsystem. As such, a holistic hardware/software approach to handling voltage emergencies has the potential to provide additional advantages beyond the fail-safe capabilities of hardware-only solutions. In particular, voltage-monitoring hardware coupled with a dynamic optimization system could be used to sense voltage emergencies, modify the problematic code sequences, and avoid future voltage emergencies in those code sequences. To be worthwhile, a holistic hardware/software approach should incur a hardware cost that is not much greater than the fail-safe circuitry currently employed and a runtime cost that is significantly less than the performance gained from avoiding future emergencies.

A runtime system can effectively balance the performance/power trade-off. Most dynamic optimizers optimize and cache the frequently executed portions of a program at the granularity of hot code traces—dynamic instruction sequences that span procedure call and branch boundaries whose code characteristics are highly tuned to the underlying hardware capabilities. Therefore, such a system can correct problems that span beyond power and performance, and into reliability. It has

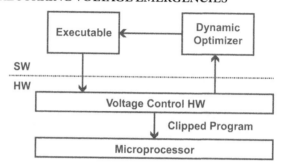

Figure 7.1: Hardware/Software Codesign. Collaborative architecture for eliminating voltage emergencies by exchanging pertinent information across traditional boundaries.

the benefit of knowing in real time when a voltage emergency occurs. Moreover, by operating in a lazy optimization mode, a dynamic optimizer can wait until it is informed by the hardware of a voltage emergency (after the hardware activates control mechanisms to eliminate the emergency), and it can then reoptimize and cache a version of the code that exhibits more voltage stability. In the ideal case, only one iteration of a power-virus loop would require hardware intervention, and the remaining iterations would be executed from the software-based, dynamically-optimized code cache.

Figure 7.1 provides a high-level view of the general software-assisted, hardware-guaranteed resilient architecture design for voltage variation. The previously proposed hardware-based voltage-control mechanisms remain intact, while the extensions are shown at the software level. The voltage-control hardware in the figure monitors execution of the application. Upon detection of an imminent voltage emergency, the control mechanism intercepts execution and performs various actions to correct the emergency. Simultaneously, the control mechanism provides feedback to the dynamic optimizer relaying pertinent information about the state of the processor during the emergency, such as the instructions that are currently in-flight or recently completed.

7.1   OPPORTUNITIES AND CHALLENGES

To begin with, the interesting research question is not whether we can build a dynamic optimizer to apply these optimizations, but whether the elimination of the identified emergency-prone activity (be it program control flow or microarchitectural activity, or a mix of both) will actually reduce the number of voltage emergencies incurred during execution. In this section, we examine whether removing an identified emergency-prone activity in targeted loops of SPEC programs helps eliminate emergencies. For example, if we identified that an L2 miss causes an emergency in a code region, then does removing that L2 miss result in that emergency not recurring? Prior work results [38] support this basic premise, but the effect of the "optimization" is not that simple and localized, as we will describe. We recommend reviewing Sections 3.3–3.4 before proceeding, to recall the relationship between microarchitectural events and program behavior and voltage emergencies.

7.1.1   OPPORTUNITIES

Let us begin by considering several programs along with the root causes of their emergencies. For instance, the top loop in *twolf* is a very small loop, nested inside a larger outer loop. The inner loop frequently suffers from branch mispredictions that lead to emergencies due to pipeline flushes. Similarly, the top loop in *apsi* has 32% of its emergencies attributed to TLB misses. In *equake*, L2 misses are the primary cause of emergency, and so forth.

In the case of *twolf*, we apply branch optimization (i.e., perfect prediction) manually to the emergency-prone branch. A hint is given to the hardware in order to enable perfect prediction. For *apsi*, all the address translations that miss in the TLB are prefetched. For loops with L2/TLB misses as the cause for emergencies, we prefetch the loads/stores that are causing misses. For loops with long latency operations, the latency of the long instructions is eliminated artificially by tweaking the simulator. For loops with several interesting events, the most recurring event is chosen and the necessary optimization is performed. These methods, done as a replacement to compiler-driven code reordering and hardware optimization, are sufficient to validate that these instructions in the code region cause the emergencies and whether they can be eliminated with the proposed techniques.

Table 7.1 shows the effect of different optimization techniques. The third column shows the effectiveness of the optimization in reducing the emergencies in that loop for each benchmark, and the fourth column represents the overall effect of the optimization in the loop on the total emergencies across the entire benchmark. This table shows that any optimization can either have an isolated effect on the loop (self-contained optimization), or have secondary effects on loops close to the optimizing loop region (spilling-over optimization). The secondary effects can be further divided into positive spill-over, in which the optimization for that particular loop altered the application's current signature so as to remove emergencies from other loops, or negative spill-over, in which the optimization removed emergencies from that loop but caused more emergencies elsewhere.

As can be seen from the Table 7.1, the optimizations are successful in removing the emergencies in the code region. The extent of success, however, varies. For example, in *equake*, where L2 misses are the primary cause of emergency, prefetching loads causing those L2 misses removes most of the emergencies. On the other hand, in *apsi*, a variety of microarchitectural events contributes to emergencies in the top loop. Hence, prefetching the TLB misses helps, but this results in less-than-optimal reduction in emergencies. In *bzip*, a change in the application's current signature as a result of removing mispredicts had a significant positive spill-over (20%) effect and resulted in other emergencies disappearing as well. However, *gzip* shows a negative spill-over, in which emergencies in the loop were reduced, but more emergencies appeared elsewhere in the code region.

In general, results indicate that directed optimization is successful. A side effect of this successful optimization points out the correctness of the categorization of the voltage emergencies. However, the presence of negative spill-over suggests that the process of optimization has to be

Table 7.1: Opportunities. Effectiveness of various optimization schemes

Benchmark	Contribution of the loop to emergencies(%)	Reduction of Emergencies in the given loop (%)	Overall Reduction (%)	Spillover (%)
Optimization: Branch Prediction				
twolf	14.9	99.9	20.9	6.0
mesa	20.9	98.8	24.3	3.6
bzip	22.1	94.9	41.4	20.4
art	80.6	79.1	68.9	5.1
mcf	31.8	77.1	68.1	43.6
gap	10.8	52.7	6.1	0.41
crafty	9.7	22.1	1.9	-0.24
gzip	61.5	19	5.4	-6.3
Optimization: Prefetching Loads				
equake	47.9	97.75	52.8	5.9
swim	71.1	80.2	57.0	-0.02
wupwise	48.2	66.6	16.0	-16
applu	23.8	32.3	7.7	0.01
gap	10.7	7.52	0.7	-0.11
Optimization: Long Latency Operations				
applu	23.8	23.4	5.6	0.03
Optimization: Prefetching TLB misses				
apsi	65.9	26.4	17.7	. 0.31

continuously carried out, perhaps throughout the application's runtime. The simplicity of the optimization mechanisms and the narrow code region that need to be altered (for example, inserting prefetch instructions before particular load instructions) make it suitable for a dynamic compiler to do them in real time without much overhead.

7.1.2   CHALLENGES

The analysis so far strongly suggests a correlation between voltage variation and microarchitectural stalls. This can lead to the false assumption of direct causality—namely, that eliminating stalls directly decreases voltage variation. Kanev et al. [51] demonstrated that the effects of compiler optimizations on voltage variation are more complex. Compiler-optimized code experiences a greater number of voltage droops, and, in certain cases, the magnitude of the droops is also noticeably larger. In a resilient design, this can eventually lead to a performance loss for the more aggressively optimized case.

Impact on Droop Counts   Typically, the task of an optimizing compiler is to increase instruction throughput through the processor. A large body of well-known optimizations (such as loop unrolling,

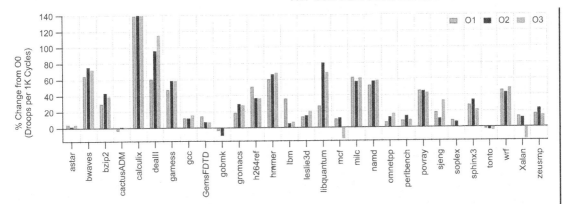

Figure 7.2: Voltage Variation Across Degrees of Compiler Optimizations. O1 includes basic optimizations (e.g., dead code elimination). O2 increases their aggressiveness without increasing code size (e.g., instruction scheduling). O3 includes the most aggressive and computationally heavy optimizations (e.g., function inlining and vectorization).

instruction scheduling, register allocation) achieves that by eliminating various microarchitectural stalls. Thus, if the analysis was interpreted as eliminating stalls would improve performance and, as a direct casualty, decrease voltage variation, one would expect that higher compiler optimization levels would decrease the amount of voltage variation, while increasing performance. However, data gathered for the GCC compiler contradicts such a notion.

Figure 7.2 shows the voltage emergency behavior of the single-core SPEC CPU2006 programs, when compiled with optimization levels ranging from O0 to O3. We classify reductions in its supply voltage below an aggressive 2.3% margin as voltage droops. We observe that increasing the aggressiveness of performance optimization with respect to the O0 baseline leads to a larger number of droops per 1K cycles in the majority of programs.

Out of the 29 runs in this experiment, 19 binaries compiled with maximum optimization result in a more than 10% increase in the number of droops, compared to the respective non-optimized versions. *454.calculix* shows the largest increase—at O3 its droop counts more than triple. The fluctuations for the other 10 programs in the experiment are predominantly smaller. Note that in several cases these small fluctuations are negative, that is, better-optimized code results in fewer droops.

Looking at the more moderate optimization levels O1 and O2 does not show a qualitative difference. The set of programs that show a large increase in droops remains largely unchanged, with the difference being in the magnitude of the increase. However, O3 does not always result in the largest variation increase, compared to O1 and O2.

The behavior of the majority of the programs is easily explicable. When better optimized at O3, programs achieve higher instruction throughput through the processor. At the microarchitectural level, this implies that pipeline utilization is high, and consequently switching factors are larger,

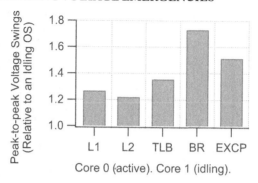

Core 0 (active). Core 1 (idling).

Figure 7.3: Measured Impact of Microarchitectural Stalls. We observe varying amount of voltage variation for different types of microarchitectural stall events.

therefore, the core consumes a relatively larger amount of current. On a stall, the net change in current is larger than in the unoptimized case. Since voltage fluctuations are proportional to such changes in current, each stall is more likely to cause a subsequent voltage droop. This effect leads to a larger aggregate number of droops over the whole execution, even though the number of stalls may be smaller.

Root Cause   In order to better understand why more aggressively optimized code can lead to a larger number of voltage droops, we must look closely at the microarchitectural foundations of droops. By using microbenchmarks, we can see that some stall events generate significantly more voltage variation than others.

Measured data on an Intel Core 2 Duo processor show that stall events contribute differently to the amount of voltage variation [80]. Figure 7.3 illustrates this with microbenchmarks consisting of simple sequences of stall events—L1 cache misses (L1), L2 cache misses (L2), data TLB misses (TLB), branch mispredictions (BR), and hardware exceptions (EXCP). The bars show the peak-to-peak magnitude of voltage swings, caused by the different stall events, normalized to the swing magnitude of the idle loop of the operating system. We can see that branch mispredictions and exceptions cause a significantly larger voltage swing than cache misses—e.g., the difference between the branch misprediction swing and that of an L2 miss is close to 50%.

There is an intuitive explanation for the results in Figure 7.3. Voltage variation is an artifact of rapid changes in processor activity. Before a miss event, the processor is executing instructions, switching factors are high, and current consumption is relatively large. A miss event throttles execution, but to a varying degree—the more severe the miss event, the larger portion of the chip stays idle to recover, hence, the lower the current consumption and the larger the voltage swing. Out-of-order pipelines are designed to mask memory misses by continuing execution, explaining the low voltage swings for L1 and L2 misses. TLB misses require more special handling (page-walking) that could keep a larger portion of the chip idle. Finally, branch mispredictions and exceptions require flushing

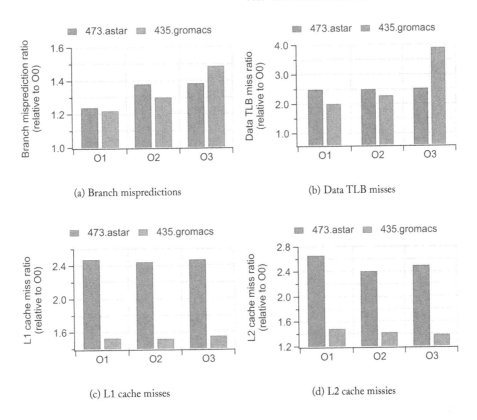

Figure 7.4: Miss Ratio and Compiler Optimization Interactions. Microarchitecture behavioral data for *473.astar* and *435.gromacs* under different compiler optimization levels.

most of the pipeline, resulting in low activity before the core pipeline is filled up, leading to a large voltage swing.

If we profile two programs as a case study, we can further demonstrate the variation-criticality of some stall events. In order to demonstrate that, Figure 7.4 shows measured miss ratios for *435.gromacs* and *473.astar*. All data in the figure are relative to the case with no optimizations (O0) and optimization aggressiveness grows to the right. For both programs, higher levels of optimization lead to higher branch misprediction and TLB miss ratios (Figure 7.4(a)-7.4(b)). Both metrics are eventually higher at O3 for *435.gromacs*, which also exhibits a large increase in voltage droops between O0 and O3 (Figure 7.2). On the other hand, both L1 and L2 miss ratios increase significantly for *473.astar* (Figure 7.4(c)-7.4(d)), without a corresponding increase in variation activity (Figure 7.2).

This data implies that in a resilient architecture design where the hardware is providing a fail-safe guarantee at some recovery penalty, the increased number of droops at higher code optimization levels has a respective net performance penalty. This penalty may be sufficiently large to even offset

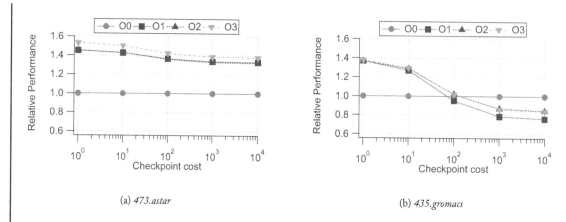

(a) *473.astar*   (b) *435.gromacs*

Figure 7.5: Impact of Compiler Optimizations. Influence of compiler optimizations on performance for varying recovery costs in a resilient processor architecture.

the initial performance gains from optimizing the code more aggressively. Therefore, we need to be aware and sensitive to this tradeoff.

Impact on Net Performance   Let us understand the tradeoff by analyzing the net performance of two programs that are representative of the general trend that Kanev et al. observed [51]. One benchmark, represented by *473.astar*, experiences little change in voltage variation across optimization. The other benchmark, represented by *435.gromacs*, experiences a significant increase in variation. Kanev et al. account for the recovery cost of each voltage emergency using a simple performance model of a resilient architecture. Namely, each crossing of the voltage margin triggers a fallback mechanism with a set checkpoint recovery cost in cycles. The cycles spent in recovery are added to the conventional running time of the program for an estimation of its runtime on a resilient architecture with the specific recovery cost.

Figure 7.5 shows performance improvement achieved by the two representative programs for different costs. Both programs rightfully receive a significant performance gain from higher levels of performance-centric compiler optimizations. For *473.astar*, this gain is sufficient to sustain higher net performance even at very large recovery costs. Even though the gains diminish because of the slightly increased droop counts (and therefore emergency recoveries), in this case, net performance is dominated by factors other than variation. For workloads represented by *435.gromacs*, fine-grained recovery presents similar results. However, after a certain recovery cost (100 cycles in this particular case), voltage emergency effects begin to dominate over the initial performance gains and less optimized binaries achieve better net performance, after factoring in emergency recovery penalties. Even the modest 30% increase in relative droop counts that *435.gromacs* shows is sufficient to offset the 50% initial performance gains from compiling with O3.

These performance results, combined with the microarchitectural root-cause analysis, strengthen the hypothesis that performance-critical and variation-critical stall events are not necessarily the same. In terms of code optimization, this suggests that purely traditional optimizations targeting, for instance only cache behavior using techniques such as software prefetching, are unlikely to fully provide a solution for voltage variation. Such differences inspire future work in finding the optimal set of code transformations for a voltage variation resilient processor architecture.

7.2 COMPILER TECHNIQUES

Currently, production compilers do not account for voltage variation when scheduling instruction sequences. Nevertheless, techniques have been developed to produce power-efficient code at the static compiler originally, and more later at the dynamic compiler level. We'll discuss both here, and explain why the dynamic compiler-based approach is likely the more suitable one to succeed in the long run to mitigate voltage variation.

7.2.1 STATIC COMPILER

Toburen [92] and Yun and Kim [98] made initial progress in the direction of compiler-driven code (re)scheduling. Toburen's approach builds an instruction schedule that limits processor power dissipation during each cycle. The power-aware scheduler attempts to place as many instructions as possible into a single VLIW instruction bundle, but within a preset power threshold. Often, high-energy instructions are not scheduled together because they can result in large and sudden current spikes. Instead, instructions are dispersed slightly from one another by exploiting scheduling slack, which is typically available if the compiler produces sufficiently large code regions. In this manner, the compiler generates a uniform $\frac{di}{dt}$ curve that decreases the processor's average peak-power consumption each cycle.

Similarly, Yun and Kim propose a power-aware modulo scheduling algorithm for high-performance VLIW processors. Their proposed algorithm reduces both the step power (the effect that causes voltage noise) and peak power by constructing a more balanced parallel schedule that does not sacrifice performance. They focus specifically on loops without conditional control transfers because it enables effective software pipelining. The "optimization" transforms a sequential loop such that new iterations start before preceding iterations finish, in order to overlap the execution of multiple iterations in a pipelined execution manner.

Software pipelining is a widely used compiler algorithm for increasing the instruction-level parallelism of cyclic code. By unrolling loops and overlapping the execution of instruction sequences from several loop iterations, the instructions can be scheduled more tightly. Typically, the result of software pipelining is that n-iterations of a loop will be combined to form one larger loop iteration. The nature of the software pipelining algorithm has two interesting side-effects. First, the technique allows high-activity periods in one loop iteration to be combined with low-activity periods of the next loop iteration, potentially leading to a more stable sequence of instructions that will often complete faster than the original sequences. Second, by changing the amount of work done in a loop iteration, periods of high and low activity that fall on the resonant frequency will be disrupted. Figure 7.6(b) depicts the result of applying software pipelining to the loop body in Figure 7.6(a). By unrolling the loop body once, and therefore lengthening the period of low activity originally resulting from three subsequent divide operations, we were able to move the stressmark off of the resonant frequency. This reduced the resulting voltage fluctuations and potentially eliminated numerous invocations of the hardware-throttling mechanism.

Even if algorithms were developed for locating potentially dangerous instruction sequences, the decision on whether to intervene would depend on the power-supply network's characteristics and the target processor's operating voltage range, which typically are not known at compile time. Finally, static techniques may not avoid all voltage emergencies, because many of the emergencies occur due to dynamic instruction sequencing, which is difficult to predict prior to program execution. Hence, Hazelwood and Brooks [40] proposed extending the hardware mechanisms to additionally provide feedback to a software-based dynamic optimization system that can determine whether a similar voltage emergency has occurred in the past, making this region of code a candidate for reoptimization.

7.2.2   DYNAMIC COMPILER

Dynamic optimization systems [8] are well-suited for emergency-specific code transformations, especially in scenarios like "90% of the execution time is spent in 10% of the code." By operating in a lazy optimization mode, the optimizer can wait until the hardware informs it of a voltage emergency (after the hardware activates control mechanisms to eliminate the emergency), and it can then reoptimize and cache a version of the code that exhibits more voltage stability. Figure 7.7 shows a typical dynamic optimizer's control flow. The system observes execution and performs code transformations to a cached copy of the frequently executed instructions. The cached, transformed code is then executed in lieu of the original code. Finally, runtime feedback and profile information is used to guide other transformations, and the process continues per emergency. Let us now examine some techniques.

Code Motion   When a static compiler schedules instructions, it often has several options for scheduling an instruction that result in equal runtime performance of the application. Thus, the compiler may inadvertently create regions of high and low processor activity simply due to its predefined settings for scheduling instructions in the event of a performance tie. By recognizing these schedule slips, a dynamic optimizer can later apply code motion to move instructions from high to low processor-utilization regions. This technique can result in the removal of a voltage emergency without degrading application performance.

Issue Rate Staggering   A dynamic compiler can slow the machine's issue rate at an appropriate point to prevent recurring emergencies altogether. The underlying assumption here is that the architecture is co-designed to signal the dynamic compiler that there is an opportunity to fix an emergency. The general idea was previously illustrated in Figure 7.1.

```
────────────── dI/dt Stressmark ──────────────
          BEFORE                      AFTER
    ldt    $f1, ($4)          ldt    $f1, ($4)
    ldt    $f2, ($6)          ldt    $f2, ($6)
    divt   $f1, $f2, $f3      divt   $f1, $f2, $f3
    divt   $f3, $f2, $f3      divt   $f3, $f2, $f3
    divt   $f1, $f2, $f3      divt   $f1, $f2, $f3
    stt    $f3, 8($4)         divt   $f3, $f2, $f4
    ldq    $7,  8($4)         divt   $f4  $f2, $f4
    cmovne $31, $7, $3        divt   $f3, $f2, $f4
    stq    $3, ($4)           stt    $f4  8($4)
    stq    $3, ($4)           ldq    $7,  8($4)
    stq    $3, ($4)           cmove  $31, $7, $3
    stq    $3, ($4)           stq    $3, ($4)
    ...                       ...
    stq    $3, ($4)           stq    $3, ($4)
```

(a) Instructions in a stressmark

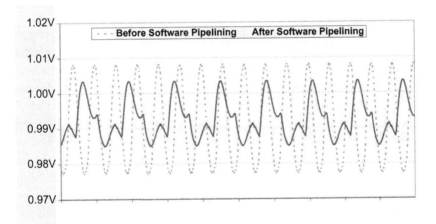

(b) Effect of software pipelining

Figure 7.6: Effect of Software Pipelining. (a) Modifying the instructions in the $\frac{di}{dt}$ stressmark (BEFORE) to a pipelined version (AFTER) reduces variation, as seen in (b).

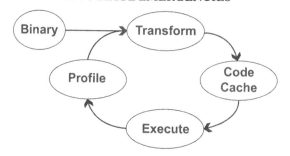

Figure 7.7: Dynamic Optimizer. Control flow of a typical runtime code optimizer [40].

One can stagger issue rate in software by altering the program code that gives rise to emergencies at execution time, and can do so without large performance penalties. The technique involves exploiting pipeline delays to decrease the issue rate close to the root-cause instruction. Pipeline delays exist because of NOP instructions or read-after-write (RAW), write-after-read (WAR), or write-after-write (WAW) dependencies between instructions. Hardware optimization techniques, such as register renaming in a superscalar machine, can optimize away WAR and WAW dependencies, so a RAW dependence is the only kind that forces the hardware to execute sequentially.

The compiler tries to exploit RAW dependencies that already exist in the program to slow the issue rate by placing dependent instructions close to one another. This constrains the burst of activity when the machine resumes execution after the stall, which prevents the emergency. Whether the compiler can successfully move instructions to create a sequence of RAW dependencies depends on whether moving the code violates either control dependencies or data dependencies. From a high level, the compiler's instruction scheduler should not break any data dependencies, and it needs to work around control dependencies by cloning the required instructions and moving them around the control flow graph such that the original program semantics are still maintained.

Figure 7.8 shows how the issue-rate smoothing technique works. The plot shows a slice of program activity corresponding to a loop within program *Sieve* from the Java Grande suite. Figure 7.8(a) shows that data dependence on a long-latency operation stalls all processor activity, so the current profile goes flat (marker 1). When the operation completes, the issue rate increases rapidly (marker 2) as several dependent instructions are successively released to functional units. This activity increases draw (marker 3), and, as a result, the voltage dips below the lower margin (marker 4). Figure 7.8(b) shows activity after the reschedule transforms the code slightly to reduce the issue rate. Because dependent instructions are packed more tightly, the issue rate in Figure 7.8(b) does not spike as high as in Figure 7.8(a) (marker 5). As a result, the processor now draws current less aggressively. The gradient at marker 6 is less steep compared to marker 3. Therefore, the original emergency at marker 4 is now permanently eliminated (marker 7).

Using this one issue-rate constraining technique, the compiler removes over 62% of all emergencies across the Java Grande suite [75]. On average, only 20% of all root causes had to be resched-

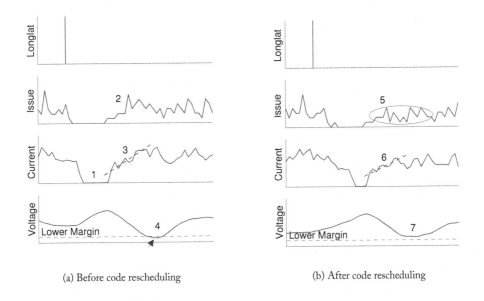

(a) Before code rescheduling   (b) After code rescheduling

Figure 7.8: A 50-cycle Execution Snapshot of *Sieve*. (a) A pipeline stall on a long latency operation triggers an emergency (indicated by an arrow) as the issue rate ramps up sharply once the operation completes. (b) Code rescheduling slows the issue rate just enough to prevent the emergency illustrated in (a).

uled because they contribute to a large percentage (over 98%) of all emergencies. These results indicate that issue-rate smoothing works well for isolated emergencies such as the case illustrated in Figure 7.8(a). However, there is a caveat. Code rescheduling works best on in-order processors where machine behavior is predictable at the compiler level. Out-of-order superscalar processors can render such compiler-level techniques ineffective because of low-level hardware-instruction scheduling. However, researchers indicated that making the RAW dependence chain as long as the issue width of the machine can overcome this hurdle effectively [76].

Instruction Padding   A final optimization is one that can be applied to acyclic regions when it is not possible to perform code motion. Instruction padding inserts unnecessary calculation into a low-utilization code region. This transformation masks the low-utilization region in a manner similar to the hardware technique of phantom firings of the functional units. Instruction padding is not used in traditional compiler-optimization phases because it has no performance benefits. Although the processor ideally will schedule the unnecessary instructions off the critical path on idle functional units, this approach may degrade the performance of an instruction sequence, and therefore should be considered as a last resort.

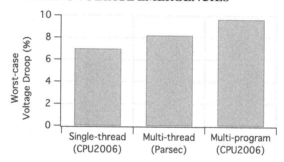

Figure 7.9: Single-thread vs. Multi-thread Impact. Worst-case droop increases during multithreaded and multiprogram execution compared to single-thread execution.

A simple way to create padding would be to insert NOP instructions into a program's dynamic instruction stream. However, modern processors discard NOP instructions at the decode stage. Therefore, the instruction does not affect the machine's issue rate. Instead of real NOPs, the compiler can generate a sequence of instructions containing RAW dependencies that have no effect. Because these *pseudo-NOP* instructions perform no useful work, this approach often degrades performance.

7.3   THREAD SCHEDULING

Simultaneous multi-threading (SMT) and other multithreading architectures have become commonplace in high-performance processors. Thus, it has become necessary to identify the temporal locality among threads (especially when that temporal locality causes recurring emergencies) and to translate that temporal locality into "simple spatial locality" that is easy for a dynamic optimizer to create, analyze, and optimize for mitigating voltage variation.

7.3.1   INTERTHREAD INTERFERENCE

El-Essay and Albonesi [27] were the first to demonstrate and mitigate the detrimental impact of multithreaded execution on voltage variation. A natural downside of SMT processors is their larger power dissipation, due to the fact that they require additional resources (e.g., registers) and that they make better use of these resources (thereby dissipating more energy) over a given period of execution. This higher power dissipation, and thus current consumption, can lead to larger current fluctuations, and thus more voltage variation. Measured data on a Core 2 Duo processor, shown in Figure 7.9, quantitatively confirms this behavior. The figure shows that the magnitude of the worst-case voltage droop is larger during multithreaded and multiprogram execution than during single-threaded execution. Therefore, thread scheduling for voltage variation is an important emerging area.

The main reason for high voltage variation when many threads are executing is because of *hoarding*. One or more threads hold resources that they release periodically, which results in sudden bursts of activity that can lead to resonance. Hoarding of resources occurs whenever a nonblocking event causes a thread to fetch and execute a large number of instructions, yet the event must complete before these instructions can be committed. This causes the thread to tie up many machine registers and issue queue slots.

Low-latency events such as L1 cache misses that hit in the L2 cache are serviced quickly to prevent significant hoarding. However, other long-latency events such as L2 cache misses can cause a large number of instructions to be bottlenecked, and subsequently cause a large bursty release. Karkhanis and Smith demonstrated that several programs in the SPEC CPU2000 integer benchmark suite are capable of executing well beyond an L2 cache miss, so far as to fill the (single-threaded) machine resources [52].

Figure 7.10 shows a hard and clear example of processor resource hoarding. Four threads hoard and periodically free a subset of the resources due to the result of a series of L2 cache misses. The different hardware threads are represented using different colors in the graph. The graphs shows the resource occupancies and events for these different threads. From the figure, we see that two of the application threads experience L2 cache misses, but, after issuing a memory request, they manage to find a significant number of independent instructions to accumulate into the machine's available resources. When the data from the L2 cache miss returns, there is a large release of dependent instructions for execution, which immediately allows other threads to issue instructions until the machine resources are once again fully occupied. These sudden bursts of activity, whenever pending memory requests are satisfied, result in large voltage variation. In the specific example, the figure also illustrates the result voltage variation, which around 300 cycles is resonating.

7.3.2   VOLTAGE SMOOTHING

We can overcome the unwanted behavior by scheduling threads intelligently. It is important to minimize emergencies in a multicore system because the power plane is typically shared across multiple cores and a droop anywhere on a common power plane can force recovery, or failure in the case of a nonresilient architecture, across all cores. When activity on one core or hardware context suddenly stalls, voltage swings, due to a sharp and large drop in current draw. However, by maintaining continuous current-drawing activity on an adjacent core also connected to the same power supply, thread scheduling dampens that current swing's magnitude. In this way, scheduling can prevent an emergency when either core or hardware context stalls [27]. Next, we will discuss ways to perform this scheduling.

Flushing   Thread-resource hoarding has been previously studied as a source of performance loss and energy inefficiency in processors. Tullsen and Brown [93] propose a method to block fetching of instructions from threads that incurred an L2 cache miss. Additionally, instructions from that thread following the L2 cache miss are all flushed from the machine to reduce resource occupancy. In effect, this allows other threads to proceed with execution, rather than being resource bottlenecked. Originally, this technique was proposed for boosting thread performance. However, El-Essay and Albonesi [27] recommend this technique to prevent thread-resource hoarding.

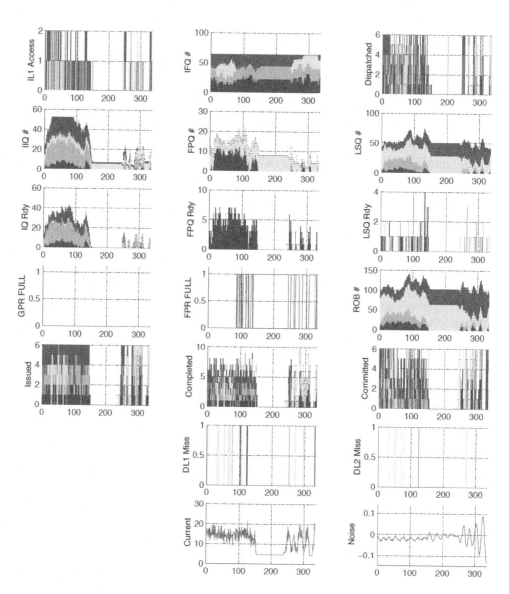

Figure 7.10: Effect of Multiple Threads. Microarchitectural activity, current, and voltage for a four-threaded workload [27].

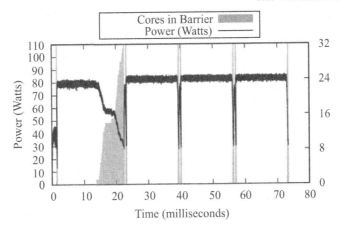

Figure 7.11: Global Synchronization Effects. Voltage variation in response to barrier synchronization for a 32-core system running the *barnes* program [63].

Similarly, El-Moursy and Albonesi [28] proposed schemes to reduce the energy consumption of issue queues in SMT processors. The general idea is to avoid idling instructions from taking up issue-queue slots by gating instruction fetch from certain threads during execution. El-Essay and Albonesi [27] believed that this technique can also be used to *proactively* mitigate voltage variation. In their research, El-Essay and Albonesi do not demonstrate a specific method that guarantees a certain level of improvement in voltage variation, but, what they do demonstrate is that multiple threads in a SMT processor can be effectively leveraged to mitigate the problem to a noticeable extent. The benefits of their approach open up ways to use a less expensive power-delivery system safely, while keeping still performance and power overheads with reasonable and acceptable levels.

Barriers   Miller et al. [63] show that any coordinated chip-wide activity in multithreaded applications on a manycore architecture can lead to much larger and rapid voltage fluctuations than in much smaller environments (e.g., a 32 core system vs. a 4 or 8 core system). One such example is barrier synchronization that is commonly used in multithreaded programs where blocked threads idle with very low power consumption. But when several idle threads are released from the global synchronization point, a sudden power spike is inevitable due the large surge in current draw.

Figure 7.11 shows the power profile for a program from the SPLASH2 application suite. Benchmark *barnes* is running on a simulated 32-core processor. *barnes* displays a very strong correlation between barrier synchronization and variation in the power profile. Here we are assuming that all the cores are connected to the same power grid. The figure shows that there is a gradual decrease in the total power consumption as cores start gathering at the first barrier. Following that, when the barrier is released, there is a sudden sharp spike in power that will likely lead to a large voltage emergency. For subsequent barriers, the threads cluster more rapidly, causing a sudden and sharp drop in the power consumption, which is immediately followed by a sharp spike in power upon release.

To mitigate the effects of barrier scheduling, Miller et al. propose a new software-level thread-scheduling technique called *VRSync* that forces more gradual and steady release of application threads from barriers, which reduces the burden on the PDN. In doing so, the amplitude of current stage is staggered over time, in effect, limiting the magnitude of the voltage variation. There are two proposed thread-release models: linear and bulk. In the linear approach, threads are released one after the other, staggered by some constant offset/delay. In the bulk mode, the technique determines the optimal number of cores that can be released all at once, but still with delays at exits.

The linear VRSync scheduler has the highest overhead. The average increases in execution time is about 11% across all the programs the authors evaluated. They find that applications that have moderate to no barrier activity have very small increases in runtime that range typically between 0 and 10%. Naturally, applications with heavy barrier activity suffer significantly more under the linear approach because more threads are forced into staggering for a longer time. For instance, they find that *streamcluster* suffered the highest overhead, with a 2.1x increase in execution time. The high overhead was because of the large number of barriers (4396) encountered in the program.

In contrast, the Bulk VRSync scheduler reduced the average execution overhead to 6.3%. The *streamcluster* program showed large improvement. Its overhead went down to 36% as compared to the 2.1x slowdown with the linear scheduler. The improvements were due to the interplay between the exit schedule and the early-exit optimization. The Bulk scheduler releases multiple threads together at the barrier exit, which quickly reach a new barrier, which naturally trigger more early exits than in the Linear case, which was the case with *streamcluster*. Thus, the Bulk scheduler is recommended and performs better.

Phases   Reddi et al. recently showed that programs experience voltage-variation phases [80] that can be overloaded with one another to smooth out voltage variation. Here, we only discuss the phase behavior. Figure 7.12 plots the droops per 1K clock cycles assuming a 4% voltage margin across three different SPEC CPU2006 programs. This metric is similar to the metric that designers typically use to study cache performance with respect to application behavior or its execution time—"misses per 1,000 instructions." The number of phases and the number of droops varies from one program to another. In Figure 7.12 program *482.sphinx* experiences nearly no phase changes. The average number of droops is stable around 10 droops per 1,000 clock cycles. In contrast, *416.gamess* experiences four phase changes where droops vary between 10 and 14 emergencies per 1,000 clock cycles transiently, while *465.tonto* goes through more complicated phase changes in Figure 7.13(c), oscillating strongly and more frequently between 4 and 12 droops every 1,000 cycles.

7.3.3   BENEFITS AND TRADEOFFS

Coscheduling threads to reduce voltage emergencies differs from scheduling for performance. This is important to understand given the large body of existing work on scheduling for better performance in multicore systems. In order to prove this point, Reddi et al. [80] evaluated different operating-

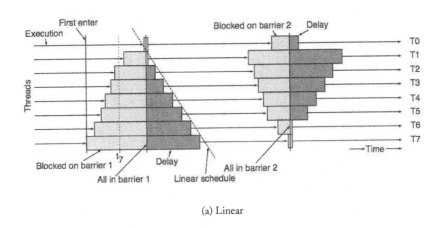

(a) Linear

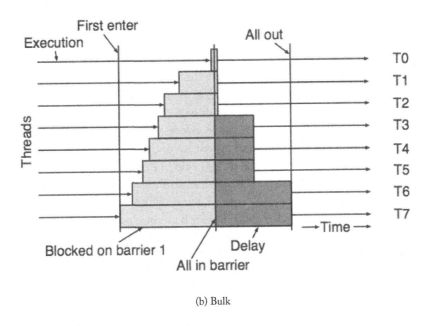

(b) Bulk

Figure 7.12: Barrier Optimization. Timing release model proposed for barrier synchronization by Miller et al. [63].

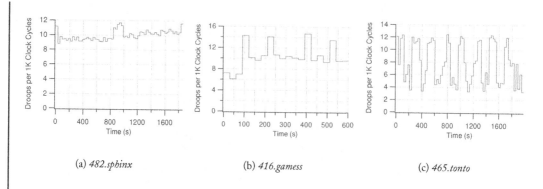

(a) *482.sphinx*   (b) *416.gamess*   (c) *465.tonto*

Figure 7.13: Voltage Noise Phases. Benchmark *482.sphinx* shows no phases, while *416.gamess* and *465.tonto* experience simple and more complex phases, respectively.

system scheduling policies, measuring emergencies over the course of a batch job schedule consisting of 50 jobs. The job pool consists of randomly chosen SPEC CPU2006 programs. Some programs may be repeatedly selected to construct the job pool, since there are only 29 CPU2006 programs. For this selected set of programs, a range of scheduling policies are also evaluated: random selection (Random) and target maximum performance (IPC), in addition to minimal emergencies (Droops).

Figure 7.14 plots performance in terms of instructions per cycle (IPC) vs. droops we observe over the course of the batch schedule. Both the y- and x-axis of the graph are normalized to SPECrate, which acts as a baseline. SPECrate assumes two instances of the same program are running together at the same time. This is done to eliminate inherent IPC differences between the programs, letting us focus in on only the effects of coscheduling. Each marker in the graph corresponds to one simulation. The experiment is conducted using 100 random simulations. The four quadrants in Figure 7.14 (Q1 through Q4) help us draw different conclusions. Ideally, we want results in quadrant Q1, which indicates that the scheduling policy lowers emergencies, in addition to improving performance. Quadrant Q2 is good, but only from a performance standpoint. Q2 suffers from an increase in emergencies. Results in Q3 are bad, since performance degrades and emergencies go up. Lastly, results in Q4 imply a reduction in emergencies at the expense of some performance.

By today's standards, the random simulation is representative of production operating systems. The POSIX 2010 policies include simple policies, such as round-robin and first-in, first-out, which are effectively random in behavior. From observing data in Figure 7.14, we can conclude that random schedules lead to more voltage emergencies. Additionally, there are no guarantees about performance.

By comparison, a performance-centric scheduler achieves best performance, as expected. However, such a scheduler is unaware of voltage-emergency activity occurring as a result of its scheduling decisions. In Figure 7.14 the IPC marker is in quadrant Q2, indicating that on aggregate more emergencies occur than the baseline. Although improving performance implicitly leads to fewer execution stalls, this data indicates that reducing stalls alone is insufficient to reduce emergencies in

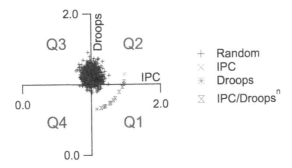

Figure 7.14: Scheduling for Voltage Variation. Scheduling for performance causes more emergencies, which upon factoring emergency-tolerance rollback costs, can actually result in performance degradation. Noise-aware schedulers are necessary in the architecture.

a multicore system. Interactions across threads (or cores) impact the amount of voltage variation we observe. Therefore, a variation-aware scheduler is necessary.

Consider the Droops metric, or voltage variation-aware scheduling, whose data point resides in Q4. The variation-aware scheduler focuses on emergency activity and therefore can minimize emergencies across all 50 jobs. It does this without adversely affecting performance. A voltage variation-aware scheduler can be adapted to not only reduce emergencies, but also, improve performance. To achieve this, a metric is necessary: $IPC/Droops^n$. Droops are weighted by some factor n that determines how costly emergencies are to tolerate. The value of n is small if recovering from emergencies is cheap, costing only few tens of clock cycles. Otherwise, n is large. A scheduler can use this value, n, to balance the penalty of tolerating emergencies as it attempts to maximize performance. The arc of markers in quadrant Q2 of Figure 7.14 illustrates the range of opportunity over different values of n.

7.4   SUMMARY

Optimizing away voltage emergencies is analogous to removing cache misses or branch mispredictions to achieve better performance and/or lower power consumption. Considering the impact of voltage variation on processor (in)efficiency, aggressive operating voltage margins are inevitable and necessary in the future. As feature-size shrinking continues, reliability problems involving voltage emergencies will emerge forcefully, requiring us to rethink traditional processor design and involving software as an essential fabric of future processor design and deployment. Collecting information about recurring emergencies and eliminating them will enable us to continue historically established reliability standards and build processors that achieve good performance within strict power and cost budgets.

CHAPTER 8

Future Directions on Resiliency

As traditional timing-margin solutions stop scaling well with reduced feature size, future systems will require adaptive processor-design techniques. The architecture must dynamically detect and recover from variation errors in the field in order to enable more-effective designs. Enabling such a system requires resiliency to be built into the system, where resiliency is a measure of a computer architecture's ability to continue working in the presence of processor degradations and failures.

In order to build a resilient processor architecture, we require a holistic system-level approach that systematically abstracts the underlying circuit-level reliability challenges to the higher levels, i.e., the architecture and software layers. The rationale is as follows. Modern applications continue to benefit from an ever-increasing amount of performance, and thus, microprocessor vendors will continue to make advances in VLSI technology, circuits, and microarchitectures to address this need. But, as we have seen, there is a growing gap between nominal operating conditions and peak operating conditions in microprocessor designs due to variations, and as such, chip manufacturers currently have two general approaches for dealing with this gap: (1) increase the cost of microprocessor-based systems by engineering them so that the hardware (including packaging and power supplies) tolerates sustained execution at the peak operating conditions, or (2) forego the costs associated with peak operating conditions and instead include hardware-based throttling mechanisms that sacrifice performance when operating conditions stray too far from nominal.

Clearly, the first approach is not an option for the commodity market, where energy efficiency and the price-to-performance ratio are more important design principles than just raw performance. The second approach seems to head in the right direction in that it provides hardware guarantees that catastrophic events will never occur. However, hardware-only solutions are reactive, lack global perspective, and may be difficult to implement efficiently. Thus, we require an approach that relies on hardware for immediate reaction (albeit suboptimal) to extreme margin violations, or emergencies, and relies on the global view provided by the software layer to eliminate repeated emergency occurrences. In this context, there are two forms of resiliency: system-level and application-level resiliency.

8.1 SYSTEM-LEVEL RESILIENCY

In this book, we discussed a paradigm of collaborative computer architecture, where both hardware and software play an integral role to diminish the detrimental effects of variations. In effect, we abstracted circuit-level reliability challenges to the higher levels, the microarchitecture and software layers. The lower layers propagated relevant information to the higher layers, as illustrated in Fig-

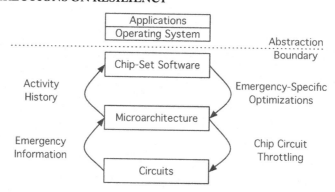

Figure 8.1: Resilient Architecture Design. An overview of exposing circuit-level challenges to the higher levels of execution.

ure 8.1. The figure presents an overview of the three levels and some of the information that flows between these levels when detecting and resolving emergencies. The bottom layer included low-level hardware and circuit blocks that signal sensed critical information about voltage emergencies. The microarchitectural layer collects this sensor data, filtering it and combining it with runtime activity history, such as the current thread and its code, along with microarchitectural state information, before passing it up to a chip-management software layer that enables transparent and error-free application-level software execution.

There are two distinct methods for dealing with emergencies in this architecture. First, the circuits and microarchitectural layer are responsible for guaranteeing reliable operation without the assistance of software. Second, the software layer seeks to eliminate these exceptional events from recurring in the future through emergency-specific dynamic optimizations. But, as a first line of defense, the circuits and microarchitectural layer operate independently of software to throttle circuit execution/behavior to guarantee correctness. An advantage of this multilayered approach is that it allows the hardware to focus on guaranteeing correct operation for the initial exceptional event, while the software focuses on eliminating or reducing the performance impact of the future events in the steady state.

Such a multilevel approach aims to eliminate the performance penalties that arise in the use of circuit techniques and microarchitectural changes that lower power, price, or, in general, attempt to optimize a design for criteria other than performance. Toward this end, we must investigate what it takes to design and build commodity computing systems that achieve both high performance and low cost today and in the future. Cost is a generic term we use as a placeholder for whatever other design criteria matters beyond performance (e.g., power, packaging costs, power supply costs, etc.). A holistic and collaborative solution enables cost-effective processor operation even in the presence of dynamic variations. By having the higher layers influence or mitigate problems at the circuit layer, the resulting costs and efficiencies help track future increases in sustained performance by maintaining the price-to-performance ratio, an important principle in the commercial sector [9].

An added benefit of system-level resiliency is transparency from the application layer. Transparency means that the programs running on the machine are unaware of the hardware emergencies that avoid the catastrophic events. System-level resiliency maintains the hardware guarantees transparently. The microcode, firmware, operating system, virtual machine layer, or compiler is notified of the emergencies handled by the hardware, and the software stack is expected to reconfigure the execution environment to avoid future occurrences of that emergency. To understand how designers can build such a transparent system-level resiliency solution, numerous questions must be answered in a systematic, cohesive, and collaborative manner. We have identified and listed a few of the most important questions in the order of bottom-up design principles.

- *What information should the circuit layer provide to the microarchitecture layer?* This involves sensing information at different points on the chip and during each quantum monitoring for emergencies. Upon detecting an emergency, the sensors must signal the processor microarchitecture of an emergency to take preventive or recovery action.

- *How should we design the microarchitecture so that it can tolerate emergencies?* Whenever we detect an emergency, the system must roll back execution to some previously known safe state. To enable this, we must understand whether we can leverage existing hardware checkpoint-recovery logic, or if we require variation-specific mechanisms.

- *What information should the microarchitecture capture?* In order to mitigate emergencies at the higher layers, the microarchitecture layer must identify the root cause of emergencies (i.e., context). Thus, it is important to identify all the microarchitectural events and types of program activity that can lead to emergencies.

- *What should the software layer look like?* Propagating architecture-level information can enable the software to eliminate emergencies. However, this requires us to identify whether a code-transforming compiler is more suitable to the problem, or whether a hardware thread scheduler is more applicable for eliminating emergencies.

- *What techniques should the software utilize to eliminate emergencies?* We must develop new heuristics and algorithms that can aid software to smooth out emergencies, be it at the level of transforming code or applying scheduling heuristics.

- *What circuit and architecture knobs are required?* Software might require hardware support to enable its optimizations, for instance to pass hints. Therefore, there needs to be a conjoined effort at the hardware and software layers so that software can explicitly drive hardware execution based on software's global knowledge and analysis.

The work done so far shows the promise of the holistic-driven approach but much remains to be done. We need a deeper understanding of how to integrate resiliency with today's workflow of complex hardware and software running on real systems. We must learn to leverage the vast work on hardware and software resiliency and develop systematic resiliency-aware design and optimization solutions that abstract hardware faults to software and integrate with the full solution. We must prototype, measure, evaluate, deploy, and revise.

8.2   APPLICATION-LEVEL RESILIENCY

Several large and important classes of applications can tolerate inaccuracies and errors in their computation. Google Search is a classic example of an application domain where transparency is less important. Google can detect and handle failures at the application level [10] because web search is inherently resilient. The global search index is split into multiple shards and these shards are serviced by multiple machines. Each of these shards is heavily replicated across numerous machines and, in the event of a failure, control flow is dynamically redirected. Other good examples of error tolerant applications are programs that come from the multimedia domain. A few bits errors in an image or a video stream may still be acceptable, and probably not noticeable at the end-user level. Some other examples include vision algorithms, machine learning, image processing, etc.

In order for application-level resiliency to succeed, we will need mechanisms that allow us to clearly express the requirements in a flexible manner that can quantitatively indicate the room for relaxing reliability constraints. For example, we need to have directives that capture the level of deadline violations (of some form or another) that are tolerable. Also, different portions of the programs may exhibit different degrees of resiliency and where and how to implement resiliency features is largely a question of efficiency that we must study.

It is worth noting that it is now commonly understood in networking that the end-user/application cannot avoid responsibility for overall correctness and reliability [84]. Decades of research and and developments in communication resilience frameworks provide us with a basis of confidence that the upper layers can systematically compensate for inherently error-prone low-level physical layer and communication reliability issues. Notable are some fascinating and counterintuitive examples where end-to-end energy costs are minimized when the resilience requirements on the lowest layers are deliberately relaxed.

Bibliography

[1] H. Akkary, R. Rajwar, and S. Srinivasan. Checkpoint processing and recovery: Towards scalable large instruction window processors. In *Proc. International Symposium on Microarchitecture, (MICRO-36)*, 2003. DOI: 10.1109/MICRO.2003.1253246 56

[2] E. Alon and M. Horowitz. Integrated regulation for energy-efficient digital circuits. In *Proc. Custom Integrated Circuits Conference*, 2007. DOI: 10.1109/JSSC.2008.925403 70

[3] H. Ando et al. A 1.3 ghz fifth-generation sparc64 microprocessor. In *Proceedings of Design Automation Conference*, 2003. DOI: 10.1109/ISSCC.2003.1234286 49

[4] H. Ando, Y. Yoshida, A. Inoue, I. Sugiyama, T. Asakawa, K. Morita, T. Muta, T. Motokuru-mada, S. Okada, H. Yamashita, Y. Satsukawa, A. Konmoto, R. Yamashita, and H. Sugiyama. A 1.3ghz fifth generation SPARC64 microprocessor. In *IEEE Journal of Solid-State Circuits*, 2003. DOI: 10.1109/ISSCC.2003.1234286 56

[5] M. Annawaram, E. Grochowski, and P. Reed. Implications of Device Timing Varaiblity on Full Chip Timing. In *Proc. International Symposium on High-Performance Computer Architecture*, 2007. DOI: 10.1109/HPCA.2007.346183 67

[6] T. M. Austin. DIVA: A reliable substrate for deep submicron microarchitecture design. In *Proc. International Symposium on Microarchitecture,*, 1999. DOI: 10.1109/MICRO.1999.809458 56

[7] K. Aygun, M. J. Hill, K. Eilert, R. Radhakrishnan, and A. Levin. Power delivery for high-performance microprocessors. In *Intel Technology Journal*, 2005. DOI: 10.1535/itj.0904.02 14, 15, 58

[8] V. Bala, E. Duesterwald, and S. Banerjia. Dynamo: a transparent dynamic optimization system. In *Proc. Programming Language Design and Implementation*, 2000. DOI: 10.1145/349299.349303 94

[9] L. A. Barroso. The price of performance: An economic case for chip multiprocessing. In *ACM Queue*, 2005. DOI: 10.1145/1095408.1095420 109

[10] L. A. Barroso, J. Dean, and U. Holzle. Web search for a planet: The Google cluster architecture. In *IEEE Micro*, 2003. DOI: 10.1109/MM.2003.1196112 110

[11] W. Becker, J. Eckhardt, R. Frech, G. Katopis, E. Klink, M. McAllister, T. McNamara, P. Muench, S. Richter, and H. Smith. Modeling, simulation, and measurement of mid-frequency simultaneous switching noise in computer systems. In *IEEE Transactions on Components, Packaging, and Manufacturing Technology, Part B: Advanced Packaging*, 1998. DOI: 10.1109/96.673703 12

[12] C. Bienia, S. Kumar, J. P. Singh, and K. Li. The parsec benchmark suite: Characterization and architectural implications. In *Proc. International Conference on Parallel Architectures and Compilation Techniques*, 2008. DOI: 10.1145/1454115.1454128 47

[13] D. Blaauw, R. Panda, and R. Chaudhry. In *Design and analysis of power distribution networks*. IEEE Press, 2001. DOI: 10.1145/1454115.1454128 12

[14] S. Borkar. Designing reliable systems from unreliable components: the challenges of transistor variability and degradation. In *IEEE Micro*, 2005. DOI: 10.1109/MM.2005.110 1

[15] S. Borkar, T. Karnik, S. Narendra, J. Tschanz, A. Keshavarzi, and V. De. Parameter variations and impact on circuits and microarchitecture. In *Proc. Design Automation Conference*, 2003. DOI: 10.1109/DAC.2003.1219020 1

[16] K. Bowman, J. Tschanz, N. S. Kim, J. Lee, C. Wilkerson, S.-L. Lu, T. Karnik, and V. De. Energy-efficient and metastability-immune resilient circuits for dynamic variation tolerance. In *IEEE Journal of Solid-State Circuits*, 2009. DOI: 10.1109/JSSC.2008.2007148 53, 65

[17] K. Bowman, J. Tschanz, S. Lu, P. Aseron, M. Khellah, A. Raychowdhury, B. Geuskens, C. Tokunaga, C. Wilkerson, T. Karnik, and V. De. A 45 nm resilient microprocessor core for dynamic variation tolerance. In *IEEE Journal of Solid-State Circuits*, 2011. DOI: 10.1109/JSSC.2010.2089657 53, 54, 55, 65, 66, 67

[18] D. Bull, S. Das, K. Shivshankar, G. Dasika, K. Flautner, and D. Blaauw. A power-efficient 32b arm isa processor using timing-error detection and correction for transient-error tolerance and adaptation to pvt variation. In *IEEE Solid-State Circuits Conference Digest of Technical Papers*, 2010. DOI: 10.1109/ISSCC.2010.5433919 65

[19] F. J. Cazorla, P. M. W. Knijnenburg, R. Sakellariou, E. Fernandez, A. Ramirez, and M. Valero. Predictable performance in smt processors: Synergy between the os and smts. In *IEEE Transactions on Computers*, 2006. DOI: 10.1109/TC.2006.108 50

[20] D. Chandra, F. Guo, S. Kim, and Y. Solihin. Predicting inter-thread cache contention on a chip multi-processor architecture. In *Proc. International Symposium on High-Performance Computer Architecture*, 2005. DOI: 10.1109/HPCA.2005.27 50

[21] H. Chen and I. Nair. Power management and its impact on power supply noise. In *Integrated Circuit and System Design. Power and Timing Modeling, Optimization and Simulation*, Lecture Notes in Computer Science. 2010. DOI: 10.1007/978-3-642-11802-9_15 45

[22] Y. Chen, K. Roy, and C.-K. Koh. Current demand balancing: a technique for minimization of current surge in high performance clock-gated microprocessors. In *IEEE Transactions on Very Large Scale Integration Systems*, 2005. DOI: 10.1109/TVLSI.2004.840404 45

[23] S. Das, D. Roberts, S. Lee, S. Pant, D. Blaauw, T. Austin, K. Flautner, and T. Mudge. A self-tuning dvs processor using delay-error detection and correction. In *IEEE Journal of Solid-State Circuits*, 2006. DOI: 10.1109/JSSC.2006.870912 53

[24] S. Das, C. Tokunaga, S. Pant, W.-H. Ma, S. Kalaiselvan, K. Lai, D. Bull, and D. Blaauw. Razor II: In Situ Error Detection and Correction for PVT and SER Tolerance. In *IEEE Journal of Solid-State Circuits*, 2009. DOI: 10.1109/JSSC.2008.2007145 53, 65

[25] S. Dighe, S. Vangal, P. Aseron, S. Kumar, T. Jacob, K. Bowman, J. Howard, J. Tschanz, V. Erraguntla, N. Borkar, V. De, and S. Borkar. Within-die variation-aware dynamic-voltage-frequency-scaling with optimal core allocation and thread hopping for the 80-core teraflops processor. In *IEEE Journal of Solid-State Circuits*, 2011. DOI: 10.1109/JSSC.2010.2080550 6

[26] A. Drake, R. Senger, H. Deogun, G. Carpenter, S. Ghiasi, T. Nguyen, N. James, M. Floyd, and V. Pokala. A distributed critical-path timing monitor for a 65nm high-performance microprocessor. In *IEEE International Solid-State Circuits Conference*, 2007. DOI: 10.1109/ISSCC.2007.373462 72

[27] W. El-Essawy and D. Albonesi. Mitigating inductive noise in SMT processors. In *Proc. International Symposium on Low Power Electronics and Design*, 2004. DOI: 10.1109/LPE.2004.1349361 15, 98, 99, 100, 101

[28] A. El-Moursy and D. Albonesi. Front-end policies for improved issue efficiency in smt processors. In *Proc. International Symposium on High-Performance Computer Architecture*, 2003. DOI: 10.1109/HPCA.2003.1183522 101

[29] D. Ernst, N. Kim, S. Das, S. Pant, R. Rao, T. Pham, K. F. C. Ziesler D. Blaauw, T. Austin, and T. Mudge. Razor: A low-power pipeline based on circuit-level timing speculation. In *Proc. International Symposium on Microarchitecture*, 2003. DOI: 10.1109/MICRO.2003.1253179 62

[30] D. Ernst, N. S. Kim, S. Das, S. Pant, R. Rao, T. Pham, C. Ziesler, D. Blaauw, T. Austin, K. Flautner, and T. Mudge. Razor: a low-power pipeline based on circuit-level timing speculation. In *Proc. International Symposium on Microarchitecture*, 2003. DOI: 10.1109/MICRO.2003.1253179 53, 65

[31] A. Fedorova. In *Operating system scheduling for chip multithreaded processors*. PhD thesis, Harvard University, 2006. 50

[32] P. Franco and E. McCluskey. On-line delay testing of digital circuits. In *Proc. IEEE VLSI Test Symposium*, 1994. DOI: 10.1109/VTEST.1994.292318 53

[33] P. Franco and E. J. McCluskey. Delay testing of digital circuits by output waveform analysis. In *Proc. IEEE International Test Conference*, 1991. DOI: 10.1109/TEST.1991.519745 53

[34] M. K. Gowan, L. L. Biro, and D. B. Jackson. Power considerations in the design of the alpha 21264 microprocessor. In *Proceedings of Design Automation Conference*, 1998. DOI: 10.1145/277044.277226 45

[35] E. Grochowski, D. Ayers, and V. Tiwari. Microarchitectural simulation and control of di/dt-induced power supply voltage variation. In *Proc. International Symposium on High-Performance Computer Architecture*, 2002. DOI: 10.1109/HPCA.2002.995694 69, 70

[36] M. Gupta, V. Reddi, G. Holloway, G.-Y. Wei, and D. Brooks. An event-guided approach to reducing voltage noise in processors. In *Proc. Design, Automation and Test in Europe*, 2009. DOI: 10.1109/DATE.2009.5090651 36, 78, 79, 82

[37] M. S. Gupta, J. L. Oatley, R. Joseph, G.-Y. Wei, and D. Brooks. Understanding voltage variations in chip multiprocessors using a distributed power-delivery network. In *Proc. Design, Automation and Testing in Europe*, 2007. DOI: 10.1109/DATE.2007.364663 14, 44

[38] M. S. Gupta, K. Rangan, M. D. Smith, G.-Y. Wei, and D. Brooks. Towards a Software Approach to Mitigate Voltage Emergencies. In *Proc. International Symposium on Low Power Electronics and Design*, 2007. DOI: 10.1145/1283780.1283808 30, 31, 34, 37, 39, 86

[39] M. S. Gupta, K. Rangan, M. D. Smith, G.-Y. Wei, and D. Brooks. DeCoR: A Delayed Commit and Rollback mechanism for handling inductive noise in processors. In *Proc. International Symposium on High-Performance Computer Architecture*, 2008. DOI: 10.1109/HPCA.2008.4658654 49, 58

[40] K. Hazelwood and D. Brooks. Eliminating voltage emergencies via microarchitectural voltage control feedback and dynamic optimization. In *Proc. International Symposium on Low Power Electronics and Design*, 2004. DOI: 10.1145/1013235.1013315 30, 36, 39, 57, 85, 94, 96

[41] R. Heald, K. Aingaran, C. Amir, M. Ang, M. Boland, P. Dixit, G. Gouldsberry, D. Greenley, J. Grinberg, J. Hart, T. Horel, W.-J. Hsu, J. Kaku, C. Kim, S. Kim, F. Klass, H. Kwan, G. Lauterbach, R. Lo, H. McIntyre, A. Mehta, D. Murata, S. Nguyen, Y.-P. Pai, S. Patel, K. Shin, K. Tam, S. Vishwanthaiah, J. Wu, G. Yee, and E. You. A third-generation sparc v9 64-b microprocessor. In *IEEE Journal of Solid-State Circuits*, 2000. DOI: 10.1109/4.881196 12

[42] D. Herell and B. Becker. Modelling of power distribution systems for high-performance microprocessors. In *IEEE Transactions on Advanced Packaging*, volume 22, 1999. DOI: 10.1109/6040.784471 15, 16

[43] D. Herrell and B. Beker. Modeling of power distribution systems for high-performance microprocessors. In *IEEE Transactions on Advanced Packaging*, 1999. DOI: 10.1109/6040.784471 12

[44] H. L. Ho, J. John E. Barth, R. Divakaruni, W. F. Ellis, J. E. Faltermeier, B. A. Anderson, S. S. Iyer, D.-K. Kim, R. W. Mann, and P. C. Parries. Low-cost deep trench decoupling capacitor device and process of manufacture. United States Patent 7,193,262, 2007. 45

[45] Intel. Intel Pentium 4 processor in the 423 pin/package /Intel 850 chipset platform, 2002. 14

[46] International Technology Roadmap for Semiconductors. In *Process integration, devices and structures*, 2002. 47

[47] International Technology Roadmap for Semiconductors. In *Process integration, devices and structures*, 2007. 5

[48] International Technology Roadmap for Semiconductors. In *Process integration, devices and structures*, 2007. 44

[49] N. James, P. Restle, J. Friedrich, B. Huott, and B. McCredie. Comparison of split-versus connected-core supplies in the POWER6 microprocessor. In *Proc. International Solid-State Circuits Conference*, 2007. DOI: 10.1109/ISSCC.2007.373412 43

[50] R. Joseph, D. Brooks, and M. Martonosi. Control techniques to eliminate voltage emergencies in high performance processors. In *Proc. International Symposium on High-Performance Computer Architecture*, 2003. DOI: 10.1109/HPCA.2003.1183526 36, 37, 69, 70, 76, 77

[51] S. Kanev, T. M. Jones, G.-Y. Wei, D. Brooks, and V. J. Reddi. Measuring code optimization impact on voltage noise. In *Workshop on Silicon Errors in Logic - System Effects*, 2013. 88, 92

[52] T. Karkhanis and J. E. Smith. A day in the life of a data cache miss. In *Workshop on Memory Performance Issues*, 2002. 99

[53] T. Karnik. Embedded error correction using resilient circuits: is this for real? In *ISAT Workshop on Resilient Computing Frameworks*, 2013. DOI: 10.1145/1687399.1687414 66

[54] Y. Kim, L. K. John, S. Pant, S. Manne, M. Schulte, W. L. Bircher, and M. S. S. Govindan. Audit: Stress testing the automatic way. In *Proc. International Symposium on Microarchitecture*, 2012. 40, 41

[55] N. Kirman, M. Kirman, M. Chaudhuri, and J. Martinez. Checkpointed early load retirement. In *Proc. International Symposium on High-Performance Computer Architecture*, 2005. DOI: 10.1109/HPCA.2005.9 49, 56

[56] R. Knauerhase, P. Brett, B. Hohlt, T. Li, and S. Hahn. Using os observations to improve performance in multicore systems. In *IEEE Micro*, 2008. DOI: 10.1109/MM.2008.48 50

[57] J. U. Knickerbocker, P. S. Andry, B. Dang, R. R. Horton, M. J. Interrante, C. S. Patel, R. J. Polastre, K. Sakuma, R. Sirdeshmukh, E. J. Sprogis, S. M. Sri-Jayantha, A. M. Stephens, A. W. Topol, C. K. Tsang, B. C. Webb, and S. L. Wright. Three-dimensional silicon integration. In *IBM Journal of Research and Development*, 2008. DOI: 10.1147/JRD.2008.5388564 45

[58] C. R. Lefurgy, A. J. Drake, M. S. Floyd, M. S. Allen-Ware, B. Brock, J. A. Tierno, and J. B. Carter. Active management of timing guardband to save energy in POWER7. In *Proc. International Symposium on Microarchitecture*, 2011. DOI: 10.1145/2155620.2155622 69

[59] M. Mack, W. Sauer, S. Swaney, and B. Mealy. IBM POWER6 reliability. In *IBM Journal of Research and Development*, 2007. DOI: 10.1147/rd.516.0763 55, 62

[60] J. Mars, N. Vachharajani, R. Hundt, and M. L. Soffa. Contention aware execution: online contention detection and response. In *Proc. International Symposium on Code Generation and Optimization*, 2010. DOI: 10.1145/1772954.1772991 50

[61] J. F. Martínez, J. Renau, M. C. Huang, M. Prvulovic, and J. Torrellas. Cherry: Checkpointed early resource recycling in out-of-order microprocessors. In *Proc. International Symposium on Microarchitecture*, 2002. DOI: 10.1109/MICRO.2002.1176234 49, 56

[62] A. A. Merchant, D. J. Sagger, and D. D. Boggs. Computer processor with a replay system. United States Patent 6,163,838, 2000. 63

[63] T. N. Miller, R. Thomas, X. Pan, and R. Teodorescu. Vrsync: characterizing and eliminating synchronization-induced voltage emergencies in many-core processors. In *Proc. International Symposium on Computer Architecture*, 2012. DOI: 10.1145/2366231.2337188 101, 103

[64] F. Mohamood, M. B. Healy, S. K. Lim, and H.-H. S. Lee. Noise-direct: A technique for power supply noise aware floorplanning using microarchitecture profiling. In *Proc. Asia and South Pacific Design Automation Conference*, 2007. DOI: 10.1109/ASPDAC.2007.358085 45

[65] S. Mukherjee, C. Weaver, J. Emer, S. Reinhardt, and T. Austin. Measuring architectural vulnerability factors. In *IEEE Micro*, 2003. DOI: 10.1109/MM.2003.1261389 56

[66] S. Mutoh, T. Matsuya, Y. Aoki, T. Shigematsu, S. Yamada, and J. Kanagawa. 1-V power supply high-speed digital circuit technology with multithreshold voltage CMOS. In *IEEE Journal of Solid-State Circuits*, 1995. DOI: 10.1109/4.400426 59

[67] S. Narayanasamy, G. Pokam, and B. Calder. BugNet: Continuously Recording Program Execution for Deterministic Replay Debugging. In *Proc. International Symposium on Computer Architecture*, 2005. DOI: 10.1145/1080695.1069994 49, 56

[68] M. Nicolaidis. Time redundancy based soft-error tolerance to rescue nanometer technologies. In *Proc. IEEE VLSI Test Symposium*, 1999. DOI: 10.1109/VTEST.1999.766651 53

[69] M. D. Pant et al. An architectural solution for the inductive noise problem due to clock-gating. In *Proc. International Symposium on Low Power Electronics and Design*, 1999. DOI: 10.1145/313817.313938 69

[70] M. D. Pant, P. Pant, and D. S. Wills. On-chip decoupling capacitor optimization using architectural level prediction. In *IEEE Transactions on Very Large Scale Integration Systems*, 2002. DOI: 10.1109/TVLSI.2002.1043335 45

[71] M. Popovich, A. Mezhiba, and E. G. Friedman. In *Power Distribution Networks with On-Chip Decoupling Capacitors*. Springer, first edition, 2007. 45

[72] M. Powell and T. N. Vijaykumar. Exploiting resonant behavior to reduce inductive noise. In *Proc. International Symposium on Computer Architecture*, 2004. DOI: 10.1145/1028176.1006726 27, 69, 70, 76, 77

[73] M. D. Powell and T. N. Vijaykumar. Pipeline damping: A microarchitectural technique to reduce inductive noise in supply voltage. In *Proc. International Symposium on Computer Architecture*, 2003. DOI: 10.1145/871656.859628 70

[74] M. D. Powell and T. N. Vijaykumar. Pipeline muffling and a priori current ramping: architectural techniques to reduce high-frequency inductive noise. In *Proc. International Symposium on Low Power Electronics and Design*, 2003. DOI: 10.1145/871506.871562 69, 70

[75] V. Reddi, M. Gupta, M. Smith, G. yeon Wei, D. Brooks, and S. Campanoni. Software-assisted hardware reliability: Abstracting circuit-level challenges to the software stack. In *Proc. Design Automation Conference*, 2009. DOI: 10.1145/1629911.1630114 39, 96

[76] V. J. Reddi, S. Campanoni, M. S. Gupta, M. D. Smith, G.-Y. Wei, D. Brooks, and K. Hazelwood. Eliminating voltage emergencies via software-guided code transformations. In *ACM Transactions on Architecture and Code Generation and Optimization*, 2010. DOI: 10.1145/1839667.1839674 97

[77] V. J. Reddi, M. Gupta, G. Holloway, M. D. Smith, G.-Y. Wei, and D. Brooks. Predicting voltage droops using recurring program and microarchitectural event activity. In *IEEE Micro*, 2010. DOI: 10.1109/MM.2010.25 83

[78] V. J. Reddi, M. S. Gupta, G. Holloway, M. Smith, G.-Y. Wei, and D. Brooks. Voltage emergency prediction: A signature-based approach to reducing voltage emergencies. In *Proc. International Symposium on High-Performance Computer Architecture*, 2009. 30, 79, 83

[79] V. J. Reddi, M. S. Gupta, K. K. Rangan, S. Campanoni, G. Holloway, M. D. Smith, G. yeon Wei, and D. Brooks. Voltage noise: Why it's bad, and what to do about it. In *Workshop on Silicon Errors in Logic - System Effects*, 2009. 44

[80] V. J. Reddi, S. Kanev, W. Kim, S. Campanoni, M. D. Smith, G.-Y. Wei, and D. Brooks. Voltage smoothing: Characterizing and mitigating voltage noise in production processors via software-guided thread scheduling. In *Proc. International Symposium on Microarchitecture*, 2010. DOI: 10.1109/MICRO.2010.35 35, 36, 40, 44, 47, 48, 90, 102

[81] A. Rogers, D. Kaplan, E. Quinnell, and B. Kwan. The core-c6 (cc6) sleep state of the amd bobcat x86 microprocessor. In *Proc. International Symposium on Low Power Electronics and Design*, 2012. DOI: 10.1145/2333660.2333745 22

[82] E. Rotenberg. AR-SMT: A Microarchitectural Approach to Fault Tolerance in Microprocessors. In *Proc. International Symposium on Fault-Tolerant Computing*, 1999. DOI: 10.1109/FTCS.1999.781037 56

[83] S. Rusu et al. A Dual-Core Multi-Threaded Xeon Processor with 16MB L3 Cache. In *IEEE Journal of Solid-State Circuits*, 2006. DOI: 10.1109/ISSCC.2006.1696062 60

[84] J. H. Saltzer, D. P. Reed, and D. D. Clark. End-to-end arguments in system design. In *ACM Transactions on Computer Systems*, 1984. DOI: 10.1145/357401.357402 110

[85] S. Shyam, K. Constantinides, S. Phadke, V. Bertacco, and T. Austin. Ultra Low-Cost Defect Protection for Microprocessor Pipelines. In *Proc. Architectural Support for Programming Languages and Operating Systems*, 2006. DOI: 10.1145/1168919.1168868 49, 56

[86] T. Slegel et al. IBM's S/390 G5 microprocessor design. In *IEEE Micro*, 1999. DOI: 10.1109/40.755464 49

[87] L. Smith, R. E. Anderson, and T. Roy. Chip-package resonance in core power supply structures for a high power microprocessor. In *Proc. International Electronic Packaging Technical Conference and Exibition*, 2001. 15, 16

[88] A. Snavely and D. M. Tullsen. Symbiotic jobscheduling for a simultaneous mutlithreading processor. In *Proc. International Symposium on Architectural Support for Programming Languages and Operating Systems*, 2000. DOI: 10.1145/378995.379244 50

[89] D. J. Sorin, M. M. K. Martin, M. D. Hill, and D. A. Wood. Fast checkpoint/recovery to support kilo-instruction speculation and hardware fault tolerance. Computing science technical report, University of Wisconsin-Madison, 2000. 49, 56

[90] L. Spainhower and T. Gregg. G4: A fault-tolerant CMOS mainframe. In *Proc. International Symposium on Fault-Tolerant Computing*, 1998. DOI: 10.1109/FTCS.1998.689495 56

[91] X. Tang, V. De, and J. Meindl. Intrinsic mosfet parameter fluctuations due to random dopant placement. In *IEEE Transactions on Very Large Scale Integration Systems*, 1997. DOI: 10.1109/92.645063 1

[92] M. Toburen. Power analysis and instruction scheduling for reduced di/dt in the execution core of high-performance microprocessors. Master's thesis, NC State University, USA, 1999. 93

[93] D. M. Tullsen and J. A. Brown. Handling long-latency loads in a simultaneous multithreading processor. In *Proc. International Symposium on Microarchitecture*, 2001. DOI: 10.1109/MICRO.2001.991129 99

[94] A. Uht. Achieving typical delays in synchronous systems via timing error toleration. Electrical and computer engineering tech report 032000-0100, University of Rhode Island, 2000. 62

[95] X. Vera, O. Unsal, and A. Gonzalez. X-pipe: An adaptive resilient microarchitecture for parameter variations. In *Proc. Workshop on Architectural Support for Gigascale Integration*, 2006. 62

[96] N. J. Wang and S. J. Patel. ReStore: Symptom-Based Soft Error Detection in Microprocessors. In *IEEE Transactions on Dependable and Secure Computing*, 2006. DOI: 10.1109/TDSC.2006.40 49, 56

[97] S. Wijeratne, N. Siddaiah, S. Mathew, M. Anders, R. Krishnamurthy, R. Anderson, J. S. Hwang, M. Ernest, and M. Nardin. A 9Ghz 65nm intel pentium 4 processor integer execution core. In *IEEE International Solid-State Circuits Conference, Digest of Technical Papers*, 2006. DOI: 10.1109/ISSCC.2006.1696066 59

[98] H.-S. Yun and J. Kim. Power-aware modulo scheduling for high-performance vliw processors. In *Proc. International Symposium on Low Power Electronics and Design*, 2001. DOI: 10.1109/LPE.2001.945369 93

[99] W. Zhao and Y. Cao. New generation of predictive technology model for sub-45nm early design exploration. In *IEEE Transactions on Electron Devices*, 2006. DOI: 10.1109/TED.2006.884077 43, 83

[100] S. Zhuravlev, S. Blagodurov, and A. Fedorova. Addressing shared resource contention in multicore processors via scheduling. In *Proc. Architectural Support for Programming Languages and Operating Systems*, 2010. 50

Authors' Biographies

VIJAY JANAPA REDDI

Vijay Janapa Reddi is an Assistant Professor in the Department of Electrical and Computer Engineering at The University of Texas in Austin. His research interests are in the area of computer systems, focusing on the interactions between hardware and software. He explores new opportunities and synergies for cross-layer solutions that improve processor- and system-level power, performance and reliability. He has co-authored over 30 papers in these areas, and has papers selected as *IEEE Micro* Top Picks and received Best Paper Awards.

Dr. Janapa Reddi has also worked in the computer industry, specifically focusing on processor architecture and compiler aspects at companies such as Intel, VMware, AMD Research, and Microsoft Research. One of his most significant contributions to the community is the Pin dynamic compiler that he co-authored on a 4-year stint at Intel. Pin is widely used in academia and industry for program introspection and analysis.

Dr. Janapa Reddi received his Ph.D. in Computer Science from Harvard University. He has a M.S. degree from the Department of Electrical and Computer Engineering at the University of Colorado at Boulder. His B.S. degree is from the Computer Engineering department at Santa Clara University.

MEETA S. GUPTA

Meeta S. Gupta is a Research Staff Member in the Reliability and Power-Aware Microarchitecture department at IBM T. J. Watson Research Center. Her research interests include high-performance computing, reliability, and power-aware computer architecture design. Dr. Gupta is involved in general areas of processor reliability, inductive noise, and Exascale systems. Microarchitectural techniques for reliability enhancement have also been part of her research focus.

Dr. Gupta received her Ph.D. degree in Electrical and Computer Engineering from Harvard University. She received a Master's in EE from the University of Southern California, Los Angeles, and a Bachelor's degree in Electrical Engineering from the Indian Institute of Technology, Delhi. Dr. Gupta also held positions in Advanced Mobile Networking Group at Lucent Bell Labs and in the High-Performance Computing group at IBM India Research Labs, Delhi. Dr. Gupta was the recipient of the IBM Ph.D. Fellowship.